3D Imaging of the Environment

This is a comprehensive, overarching, interdisciplinary book and a valuable contribution to a unified view of visualisation, imaging, and mapping. It covers a variety of modern techniques, across an array of spatial scales, with examples of how to map, monitor, and visualise the world in which we live. The authors give detailed explanations of the techniques used to map and monitor the built and natural environment and how that data, collected from a wide range of scales and cost options, is translated into an image or visual experience. It is written in a way that successfully reaches technical, professional, and academic readers alike, particularly geographers, architects, geologists, and planners.

FEATURES

- Includes in-depth discussion on 3D image processing and modeling
- Focuses on the 3D application of remote sensing, including LiDAR and digital photography acquired by UAS and terrestrial techniques
- Introduces a broad range of data collection techniques and visualisation methods
- Includes contributions from outstanding experts and interdisciplinary teams involved in earth sciences
- Presents an open access chapter about the EU-funded CHERISH Project, detailing the development of a toolkit for the 3D documentation and analysis of the combined coastline shared between Ireland and Wales

Intended for those with a background in the technology involved with imaging and mapping, the contributions shared in this book introduce readers to new and emerging 3D imaging tools and programs.

3D Imaging of the Environment

Mapping and Monitoring

Edited by John Meneely

CRC Press
Taylor & Francis Group
Boca Raton London New York

CRC Press is an imprint of the
Taylor & Francis Group, an **informa** business

Designed cover image: © Historic Environment Scotland, The Engine Shed, Stirling, Scotland

First edition published 2024
by CRC Press
2385 NW Executive Center Drive, Suite 320, Boca Raton FL 33431

and by CRC Press
4 Park Square, Milton Park, Abingdon, Oxon, OX14 4RN

CRC Press is an imprint of Taylor & Francis Group, LLC

ISBN: 978-0-367-33793-3 (hbk)
ISBN: 978-1-032-10895-7 (pbk)
ISBN: 978-0-429-32757-5 (ebk)

DOI: 10.1201/9780429327575

Typeset in Times
by Apex CoVantage, LLC

To my wife Dr. Julie Meneely, thank you for your never-ending support.

Contents

Preface

The field of 3D imaging has undergone a remarkable transformation over the past few decades, fuelled primarily by the emergence of new technologies, the growing demand for realistic, immersive visual experiences and the ability to easily share the results online.

This text is an essential guide for anyone interested in learning about this exciting field. Whether you are a student, a researcher, or a professional, this book provides a wide-ranging overview of the techniques, technology, and applications used today.

With contributions from leading specialists in the field, it covers an expansive range of topics and technologies over a vast range of scales, such as photogrammetry, laser scanning, drone mapping, and 3D printing. Through a collection of case studies, it explores the applications of 3D imaging in various fields, such as our built cultural heritage, geomorphology, archaeology, zoology, and climate change and how 3D technologies are being used to map, monitor, visualise, and share the research being carried out in these areas.

The text is designed to be accessible to a broad spectrum of readers, from beginners to advanced users, and includes many links to online 3D content of the examples covered. It provides a general introduction to the field of 3D imaging while also giving in-depth coverage of advanced 3D technologies and techniques.

I hope that this book will serve as a valuable resource for anyone interested in 3D imaging and that it will inspire new ideas and innovations in this exciting and rapidly evolving field.

John Meneely

Editor

John Meneely is the founder of 3D Surveying Ltd, having previously worked as Senior Research Technician at the School of Natural and Built Environment, Queen's University, Belfast. With over 30 years of experience in practical research, he has worked all over the world with interdisciplinary teams across the earth sciences. His expertise lies in using a variety of 3D laser scanning and other digital technologies to map, monitor, and visualise the built and natural environment across a wide range of spatial and temporal scales. He has presented his work at many national and international conferences and been the keynote speaker at several 3D digital technologies conferences. He was on the advisory board for SPAR Europe for two years – Europe's largest 3D scanning conference – and invited to speak at the 2009 International Council on Monuments and Sites (ICOMOS) symposium in Malta on the use of terrestrial laser scanning. His early research and publications focused on studying the catastrophic decay of building stone under complex environmental regimes and the digital documentation of natural and built heritage sites for several geological, geographical, archaeological, managerial, and educational applications. His recent interest has extended his data collection skills into 3D visualisation via 3D printing, VR, and AR. He is currently advising several SMEs, primarily in the environmental monitoring, built heritage, construction, and facilities management sector on integrating 3D technologies into their workflow.

Contributors

Kathryn Adamson
Department of Geography
Manchester Metropolitan University
Manchester, England, UK

Louise Barker
Department of Archaeology
Royal Commission on the Ancient
    and Historic Monuments of Wales
    (RCAHMW)
Shrewsbury, England, UK

Iestyn D. Barr
Department of Geography
Manchester Metropolitan University
Manchester, England, UK

James Barry
Geological Survey Ireland
Department of the Environment,
    Climate and Communications
Dublin, Ireland

Michela Bertolotto
School of Computer Science
University College
Dublin, Ireland

Marco Callieri
Visual Computing Laboratory
Institute of Information Science and
    Technologies
National Research Council of Italy
Pisa, Italy

Michael Casey
Dublin, Ireland

Paul Chapman
School of Simulation and Visualisation
Glasgow School of Art
Glasgow, Scotland, UK

Kendrew Colhoun Director
KRC Ecological Ltd
Newcastle, Northern Ireland, UK

Anthony Corns
The Discovery Programme
Dublin, Ireland

Kieran Craven
Geological Survey Ireland
Department of the Environment,
    Climate and Communications
Dublin, Ireland

Sean Cullen
Geological Survey Ireland
Department of the Environment,
    Climate and Communications
Dublin, Ireland

Sarah Davies
Department of Geography and Earth
    Sciences
Aberystwyth University
Aberystwyth, Wales, UK

Colm Donnelly
School of Natural and Built
    Environment
Queen's University
Belfast, Northern Ireland, UK

Toby Driver
Department of Archaeology
Royal Commission on the Ancient
    and Historic Monuments of Wales
    (RCAHMW)
Shrewsbury, England, UK

Trevor Fisher
Banbridge, Northern Ireland, UK

Adam Frost
Historic Environment Scotland
The Engine Shed
Stirling, Scotland, UK

Michal Gallay
Institute of Geography
Faculty of Science
Pavol Jozef Šafárik University
Košice, Slovakia

Tim Gomes
Idaho Virtualization Laboratory
Idaho Museum of Natural History
Idaho State University
Pocatello, Idaho, USA

Hywel Griffiths
Department of Geography and Earth
    Sciences
Aberystwyth University
Aberystwyth, Wales, UK

Harkin Aerial
Oyster Bay, New York, USA

Scott Harrigan
Virtual Surveyor
Raleigh, North Carolina, USA

Sandra Henry
The Discovery Programme
Dublin, Ireland

Zdenko Hochmuth
Institute of Geography
Faculty of Science
Pavol Jozef Šafárik University
Košice, Slovakia

Markus Hollaus
Research Group Photogrammetry
Department of Geodesy and
    Geoinformation
Vienna University of Technology (TU
    Wien)
Vienna, Austria

Daniel Hunt
Department of Archaeology
Royal Commission on the Ancient
    and Historic Monuments of Wales
    (RCAHMW)
Shrewsbury, England, UK

Benedikt Imbach
Aeroscout GmbH
Hochdorf, Switzerland

Stuart Jeffrey
School of Simulation and Visualisation
Glasgow School of Art
Glasgow, Scotland, UK

Sarah Kandrot
Department of Geography
University College Cork
Cork, Ireland

Ján Kaňuk
Institute of Geography
Faculty of Science
Pavol Jozef Šafárik University
Košice, Slovakia

Michael King
Newry, Mourne, and Down District
    Council
Downpatrick and Newry, Northern
    Ireland, UK

Debra Laefer
Tandon School of Engineering
New York University
Brooklyn, New York, USA

Timothy Lane
Department of Geography
    and Environmental
    Science
John Moores University
Liverpool, England, UK

Alan Lauder
Wildlife Conservation and Science Ltd
Kilcoole, Wicklow, Ireland

Ciaran Lavelle
National Museums Northern Ireland
Cultra, Northern Ireland, UK

Tony Martin
School of Natural and Built
  Environment
Queen's University
Belfast, Northern Ireland, UK

John Meneely
3D Surveying Ltd
Banbridge, Northern Ireland, UK

Aaron Miller
School of Natural and Built Environment
Queen's University
Belfast, Northern Ireland, UK

Daniel Moloney
Redcastle, Donegal, Ireland

Konstantin Nebel
Geography and Environmental
  Research Group
John Moores University
Liverpool, England, UK

Ulrich Ofterdinger
School of Natural and Built Environment
Queen's University
Belfast, Northern Ireland, UK

Edward Pollard
The Discovery Programme
Dublin, Ireland

Victor Portela
School of Simulation and
  Visualisation
Glasgow School of Art
Glasgow, Scotland, UK

Jesse Pruitt
Idaho Virtualization Laboratory
Idaho Museum of Natural History
Idaho State University
Pocatello, Idaho, USA

Patrick Robson
Department of Geography and Earth
  Sciences
Aberystwyth University
Aberystwyth, Wales, UK

Ján Šašak
Institute of Geography
Faculty of Science
Pavol Jozef Šafárik University
Košice, Slovakia

Robert Shaw
The Discovery Programme
Dublin, Ireland

Linda Shine
The Discovery Programme
Dublin, Ireland

Brian Sloan
School of Natural and Built
  Environment
Queen's University
Belfast, Northern Ireland, UK

Jozef Šupinský
Institute of Geography
Faculty of Science
Pavol Jozef Šafárik University
Košice, Slovakia

Leif Tapanila
Idaho Virtualization Laboratory
Idaho Museum of Natural
  History
Idaho State University
Pocatello, Idaho, USA

Willem G. M. van der Bilt
University of Bergen
Bergen, Norway

Evelyn Vollmer
Idaho Virtualization Laboratory
Idaho Museum of Natural History
Idaho State University
Pocatello, Idaho, USA

Anh Vu Vo
School of Computer Science
University College Dublin, Ireland

Stephen Weir
National Museums Northern
    Ireland
Cultra, Northern Ireland, UK

Lyn Wilson
Historic Environment Scotland
The Engine Shed
Stirling, Scotland, UK

Carlo Zgraggen
Aeroscout GmbH
Hochdorf, Switzerland

1 Digital Documentation and Digital Innovation in Practice
The Historic Environment Scotland Approach

Adam Frost and Lyn Wilson

INTRODUCTION AND BACKGROUND

Historic Environment Scotland (HES) (and its predecessor organisation Historic Scotland) has used digital documentation and visualisation in various forms for many years, primarily for conservation purposes. As an early adopter of these technologies, we first commissioned stereo pair photogrammetry in 1974 to augment our survey record for monitoring the condition of the historic sites we look after around Scotland, including Edinburgh and Stirling Castles. Laser scans were first commissioned in 2001 to create accurate 3D records of Pictish carved stones. In-house skill in digital documentation was established in 2007 with a specific focus on conservation. Within our Conservation Science Team, as the emphasis on non-destructive analytical techniques naturally increased, we developed our capabilities in laser scanning and infrared thermography, amongst other methods. Our expertise grew from this solid scientific foundation.

In 2009, we set out to lead the ambitious Scottish Ten Project to create accurate 3D models of Scotland's (as then) five UNESCO World Heritage Sites and five international heritage sites.[1] Initiated and funded by the Scottish Government and delivered in partnership with The Glasgow School of Art, and the nonprofit organization, Cyark, the models were used in conservation and management, interpretation and virtual access (Wilson et al., 2013). Key achievements of the project included our digital documentation work contributing to two UNESCO World Heritage Site nominations for Rani ki Vav in Gujarat, India, in 2014 and the Giant Cantilever Crane in Nagasaki, Japan, in 2015. Additionally, our accurate 3D model of Sydney Opera House was the first as-built record of the structure and has since been used to develop a detailed building information model. Within Scotland, our project captured baseline 3D data that now allows us to regularly monitor coastal erosion at Skara Brae in Orkney and to develop a city scale model for the entire World Heritage area of the Old and New Towns of Edinburgh, now often used for town planning purposes.

DOI: 10.1201/9780429327575-1

A dedicated Digital Documentation Team was established in 2010 with a focus on conservation-driven projects within Historic Scotland and collaborative working with the Visualisation Team at The Glasgow School of Art and others in the heritage sector. The experience gained delivering the Scottish Ten project and the growing appreciation that 3D data could be used for many different heritage purposes led to the announcement of the Rae Project in late 2011. The Rae Project aims to digitally document in 3D all 336 properties under the care of HES and the 41,000 associated collections items (Hepher et al., 2016). In doing so, Scotland aims to become the first country in the world to digitally document its most significant heritage assets. The project is named after John Rae, the 19th-century Orcadian explorer credited with discovering the Northwest Passage—a navigable Arctic route from the Atlantic Ocean to the Pacific. While the 3D data collected is still used principally for conservation, multiple applications in facilities management, interpretation, learning and research are increasingly common. Given the significant time, skill and financial investment in capturing and processing 3D data for complex historic sites and objects, it is eminently sensible to maximise re-use of the data sets for as wide a range of applications as possible. At the time of writing, we have completed digital documentation of approximately 42% of the properties in our care and several hundred collections items. A prioritization matrix with input from across HES business areas helps us strategically manage our ongoing programme.

In 2017, our commitment to the application of digital technologies for the benefit of our historic environment was extended with the establishment of our Digital Innovation Team. This strategic development aligned with the opening of the Engine Shed,[2] Scotland's building conservation centre, where the Digital Documentation and Digital Innovation Teams are based. The Digital Innovation Team's role is to develop the 3D data the Digital Documentation Team creates and use this in new and innovative ways, both for use by the heritage sector and beyond. Another part of the team's role is looking for cutting-edge developments in digital technologies that can be adopted for positive impact within HES. With all our digital heritage activities, we strive for high quality, maximum benefit and adherence to best practice.

OUR APPROACH

The Digital Documentation and Digital Innovation Teams work within the Conservation Directorate at HES to promote, develop and apply digital technologies to improve how we care for our cultural heritage. We aim for tangible, applied uses of the technology and data to serve as our use-cases and represent process efficiencies—where time and energy overheads may be reduced with improved results and fewer risks. This includes surprising and novel uses of emerging digital techniques, such as to support traditional skills by using additive manufacture of carved stone elements as sophisticated reference material for trained stone masons. This section looks at our general approach and highlights our ambitions for the future of digital documentation.

STRATEGIC CONTEXT

The historic environment is defined by Our Place in Time: The Historic Environment Strategy for Scotland (OPiT) (HES, 2014) as 'the physical evidence for human activity that connects people with place, linked with the associations we can see, feel and understand'. Therefore, the historic environment is both itself a collection—of buildings, landscapes and sites—and the home for diverse collections—objects, material culture and intangible heritage, stories, and traditions—that help us to understand it.

As the lead public body for the historic environment in Scotland, we work with and support the sector to deliver the aims of OPiT. Our vision is for the historic environment to be cherished, understood, shared and enjoyed with pride, by everyone. To fulfil this vision, our Corporate Plan, Heritage For All (HES, 2019b) outlines key outcomes that we will achieve through our work.

The preservation and management of our collective heritage is a key focus of our Corporate Plan. Both the properties in care and associated collections undergo continuous monitoring and targeted analysis which informs, and is informed by, our research programmes. Our ability to care for, manage, conserve and provide access to—and engagement with—those collections is underpinned by our understanding of the material science of the objects themselves and the impact that their environment has upon them. This environment—whether it is the historic buildings that house them, the impact of climate change, or the need for interventions such as heating, humidification and other forms of protection—is determined by our research and development. This, in turn, leads to policy approaches and guidance which we share with the sector. Research directly supports delivery of the HES Climate Action Plan, Research Strategy, and Digital Strategy (HES, 2019a, b & c and 2020) and broader objectives of the HES Corporate Plan.

At a wider strategic level, HES is part of the National Heritage Science Forum member council, which supports the development of heritage science infrastructure across the UK, including provision of access to specialist equipment across the membership providing a UK-scale heritage science network. The multiple value of our digital work is recognised by UNESCO and is featured strongly in the organisation's 2019 'Cultural Heritage Innovation' report (UNESCO, 2019).

RESEARCH AND SKILLS DEVELOPMENT

The HES digital teams are actively involved in research and development, both as internal R&D projects and through collaborative partnerships with higher education institutions. We support master's degree programmes and co-supervise PhD students. Research must have an applied heritage focus with clear, practical benefits. Our research activities are crucial for maintaining a leading edge in the rapidly developing area of digital heritage. Building others' capacity in digital documentation is also a focus area. Since 2012, we have hosted a funded trainee within our team for one year, offering the opportunity for hands-on, real-world experience. We also provide advice to the heritage sector in Scotland, the UK and often internationally.

One example of our collaborative research is the development of algorithms based on signal processing to automatically segment masonry walls from 3D point clouds (Valero et al., 2018 and 2019), a task that often would be accomplished by hand with inconsistent results. This research, with The University of Edinburgh and Heriot Watt University, specifically addresses maintenance of traditional buildings where elevations are produced that include the delineation of each stone and

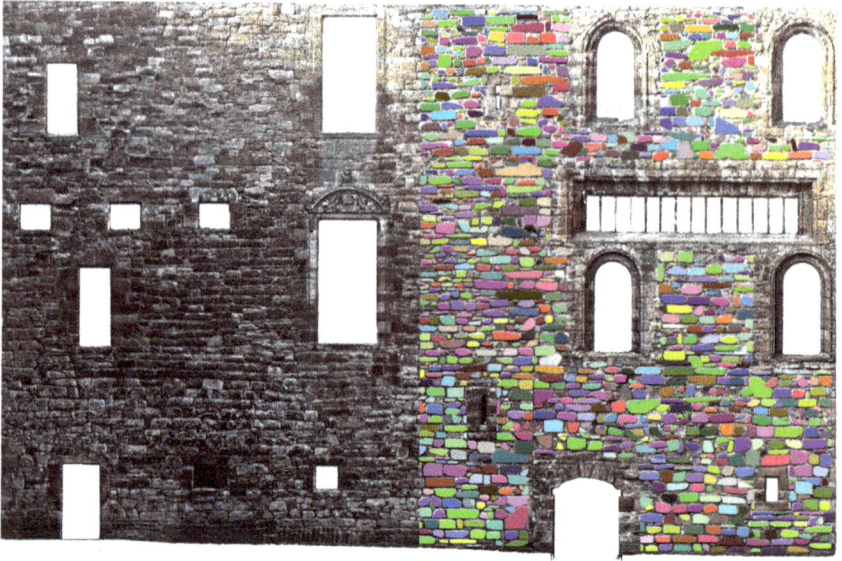

FIGURE 1.1 Automatic segmentation of rubble masonry wall at Linlithgow Palace from 3D point cloud data.

Source: © The University of Edinburgh, Heriot Watt University and Historic Environment Scotland

defect assessment, a process that may require the erection of costly scaffold. The algorithms can automatically extract this information on stone from the point cloud (Figure 1.1), potentially representing significant time and money savings. In the case of rubble masonry, the algorithm can also calculate mortar recess. The outcome of this research has been freely disseminated as an official plugin for CloudCompare[3] for use within the sector and by practitioners and specialists.

DISSEMINATION

Sharing our data in various formats allows us to facilitate research, support conservation objectives and engage wider audiences. This is central to how our work and the digital assets created through digital documentation provide value for the organisation and partners. There are technical challenges to making substantial 3D data available to end users, and we are committed to using a range of technologies and approaches best suited to the output requirements and audience. Our current approach is illustrated in Table 1.1, where our deliverable data is divided into two categories of access, each with different audiences and purposes.

Data that is required for internal organisational use, such as supporting a new architectural design in a CAD or BIM software package or an engineering enquiry, may be best used as a 'raw' or optimised interoperable 3D file. Other conservation tasks, such as condition inspection, may benefit more from visual interrogation of a 3D mesh with an emphasis on high-quality photo textures. In this case, the data

TABLE 1.1

Various Dissemination Channels for 3D Digital Documentation Data Used by the Digital Documentation and Innovation Teams within the Organisation and Externally

Access Medium	Organisational Use Only	Organisational and External Use
Network streaming	• ContextCapture Viewer (*3D textured mesh/model viewer*) • Potree Viewer (*3D point cloud viewer*) • 3DHOP Viewer (*3D viewer*)	• Sketchfab
Offline and local data access	• Leica TruView (*Enhanced panoramic laser scan viewer*) • Interoperable 3D file format (e.g., e57)	• Interoperable 2D/3D file formats (for research applications and requests)

may be better used via a network-accessible model viewer. To improve discoverability of PIC-related information within the organisation, HES has developed the Properties in Care Asset Management System (PICAMS). PICAMS serves information across the organisation via a web-accessible intranet, using enhanced search tools and geographic information to identify and present relevant information for staff enquiries. We are in the process of integrating web-accessible 3D assets with the aim of enabling organisation-wide access to Rae Project data.

For wider public dissemination (and also organisational use), further considerations that we take into account include:

- Digital access, including discoverability, availability, controls and accessibility
- Bandwidth and system resources (such as memory and graphics capability) as limiting factors to users with diverse mobile and desktop hardware
- Ensuring a good user experience; optimising 3D data, the environment and narrative elements such as annotations to engage audiences

Since 2017, we have used the online 3D model platform Sketchfab to make available and promote optimised versions of our data for sites and objects. This allows wider public access to our digital documentation data to engage audiences at different levels. As of December 2020, 354 3D models are available to view and interact with. The various models on the HES account have received over 168,000 views, 658 likes and a number of 'Staff Picks'. Figure 1.2 shows an example of an optimised 3D model of Caerlaverock Castle, presented with a virtual tour of annotated signage.

DIGIDOC CONFERENCES

The DigiDoc conference series aims to inspire and engage audiences in the application of innovative technologies for cultural heritage. First held in 2008, when digital

FIGURE 1.2 (A) Caerlaverock Castle available to view and explore for free on the HES Sketchfab account. (B) Interactive annotations within the Sketchfab model act as a virtual tour to online visitors.

documentation in cultural heritage was still in its infancy, the conference returned in 2009, 2011 and 2012. In October 2018, we welcomed international delegates to the Engine Shed in Stirling for a revival of our DigiDoc conference series. Over 200 participants from Scotland, UK, Europe, USA, China and Australia attended over three days. The Engine Shed provided the perfect venue: our dynamic building conservation centre which we aim to develop into an international centre of excellence for heritage innovation.

Our 2018 DigiDoc Research and Innovation Day provided a friendly platform for researchers and SMEs to showcase new and cutting-edge work, along with established leaders from the UK digital heritage sector. The following two days of DigiDoc saw world-leading innovators from gaming, TV, film and VFX industries and heritage, arts, science and education sectors give keynote addresses on a diverse range of thoroughly engaging topics. HES now aims to host DigiDoc on a three-yearly basis, highlighting the most inspirational and aspirational developments and applications in digital technologies.

SUSTAINABILITY

Climate change is leading to unprecedented impacts on Scotland's historic environment. As outlined in our new Climate Action Plan (CAP) 2020–2025 (HES, 2020), rising sea levels, higher temperatures and a wetter climate are pushing many historic buildings and sites into new and uncharted conditions that they were not designed to cope with. To address both the impacts and causes of climate change requires swift and meaningful action: both to reduce our own carbon footprint and to adapt Scotland's historic environment to meet the consequences of climate impacts. The CAP sets out our priorities for addressing climate change, both as an organisation and on behalf of Scotland's historic environment sector. Research and innovation is specified as one of four cross-cutting priorities underpinning delivery of this plan, and in alignment with our Research Strategy, we identify specific programmes of research activity each year to support delivery of the Plan.

As an organisation, in the last 10 years, we have reduced emissions from HES operations by 37%, and our CAP sets out not only our own net-zero carbon ambitions but our commitment to support decarbonisation and enhanced sustainability across the heritage sector in Scotland and beyond. Helping the sector to reach net-zero, using our historic assets to create positive environmental impacts and setting best practice standards through our research, education and training programmes, directly aligns with the values of the UKRI Sustainability Strategy. We are starting to explore ways to identify carbon reduction made through digital co-working, where the use of 3D digital documentation data shared with colleagues and partners is helping to reduce unnecessary travel and associated energy and emissions for projects.

METHODOLOGY

Our general digital documentation methodology is presented here, with a focus on how we tackle the differences in scale between the objects, sites and the landscapes they sit within. The Rae Project largely governs our approach to this process, with a specification that has evolved to reflect what we believe to constitute reliable, accurate 3D data for a range of uses. We also factor in enquiries relating to the properties or collections, for example, whether there is a specific research objective or application.

We work with colleagues across our organisation to facilitate access and coordinate all data capture activities for digital documentation projects. As a significant undertaking with a long-term vision, the Rae Project is structured with respect to organisational priorities. For properties, this is governed by a prioritisation matrix that considers a range of factors including conservation conditions and strategic value—where the data will be of immediate benefit to current or future investment projects, interpretation design, etc. Likewise, close liaison with the collections team allows us to digitally document items considered to be the most relevant priority candidates based on a range of criteria.

DIGITAL DOCUMENTATION AND VISUALISATION WORKFLOW

Working as part of a team is an essential element to our coordination and execution of a successful fieldwork project. From before the planning stage, through to on-site fieldwork and data management, good communication within the team is crucial. This ensures that coverage is as complete as possible, that the work is done safely, that laser scans and photography are well integrated and that relevant paradata and metadata are documented and backed up.

From a data processing perspective, the specific workflows for each digital documentation process are different, largely depending on the capture devices, associated software, data types and interoperability with other data. Our general approach to these workflows is to make use of parallel processes to save time (both staff and computational resource), due in part to the overwhelming data management and processing tasks involved that can range from fully supervised, to 'one-click' unsupervised operations.

Diagram 1.1 provides an illustrative overview of our digital documentation methodology, showing encompassing stages from planning through to the dissemination of results.

DATA CAPTURE

We use established and emerging geospatial and imaging technologies to digitally document cultural heritage assets. Through the course of the Rae Project and through collaborative projects, the technologies and workflows employed by the Digital Documentation and Innovation Teams have evolved significantly. These are rapidly developing and continue to drive improvements to our data sets, from more complete levels of coverage to higher quality 3D and image data. Speed improvements to the capture techniques and other connected parts of the pipeline (e.g., storage devices, network connectivity) allow faster on-site and on-premises data capture, saving staff time and accelerating project progress.

In this section we will look at scale as the governing factor that informs the technique and methodology that we employ to digitally document an asset. Subjects can vary massively in scope from great landscapes of cultural significance to millimetre-size collections items. Despite the unique challenges of working across these scales, there are common themes between our best practices for data capture:

- Using techniques with metric tolerances appropriate to the subject scale to generate reliable 3D data sets (accuracy, precision and resolution)
- Suitability for use in a conservation context (non-destructive, non-contact)
- Accurately reflects the condition of the subject at time of the data capture
- Reproducible methodology to monitor condition
- Conducting work safely, with regards to the safety of the survey staff, members of the public and other stakeholders
- Producing data that can either be used directly by, or adapted for end users, who may have varied technical ability and access requirements

Planning & advanced work

| Risk Assessment & Method Statement | Installation of Permanent Survey Markers (PSM) | UAV pre-flight planning and permissions |

Control survey

| Total Station survey | GNSS survey of PSM stations |

Primary spatial & image data capture

| Terrestrial laser scanning | Photogrammetry (Terrestrial) | Photogrammetry (Aerial) |

| Data filtering, colourisation, import |

Data processing and registration

| Laser scan registration | Image processing, colour correction, format conversion |

| Data export | Photogrammetric alignment and processing |

Archival and output dataset

| Archival version |

Dissemination

Public

| Retopology for real-time web 3D | Pre-rendered visualisations | VR and AR applications |

Professional

| Research specific formats, raw and processed | Architectural or engineering format | Conversion to network streamable 3D format |

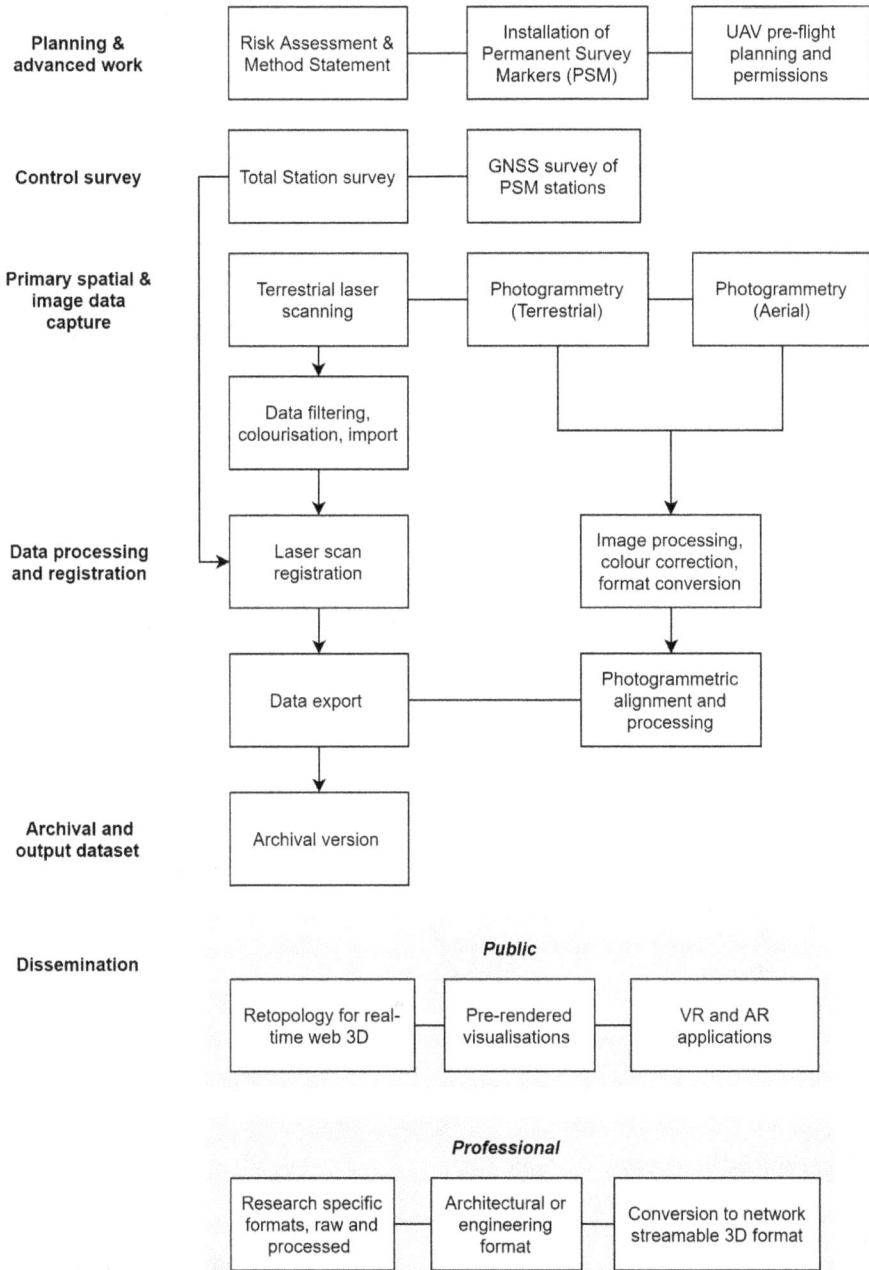

DIAGRAM 1.1 Illustrative diagram showing the HES digital documentation workflow with key outputs.

In addition to cultural heritage assets, the Digital Innovation Team at HES explores the use of current and emerging techniques to document environments to produce immersive data sets and create engaging content. For audiovisual data capture where the subject or environment may be a live event or have a temporal component, we employ some of the following techniques:

- Spatial (ambisonic) audio
- Stereoscopic video
- 360° spherical panoramic images

These capture technologies directly feed into immersive content pipelines, enabling the creation of more engaging content for public dissemination, richer digital inter-pretation and more informative learning resources for training applications. This includes both three-degrees-of-freedom (3DOF) and six-degrees-of-freedom (6DOF) Virtual Reality (VR), Augmented Reality (AR) and other classes of digital multime-dia and real-time experiences.

MULTI-SCALAR APPROACH TO DIGITAL DOCUMENTATION

We take a multi-scalar approach to digital documentation, from small artefacts to architectural features, buildings and engineering structures to entire landscapes. This variation between methodologies, software and workflows requires best practice to be employed at all levels to produce robust and reliable results. Furthermore, many of HES' 336 properties in care include natural and built heritage often spanning thousands of years of history. These sites vary from small remote monuments such as the glacial erratic Wren's Egg through to Edinburgh Castle within the UNESCO Old and New Towns of Edinburgh World Heritage Site. Our work with external partners such as Transport Scotland and Network Rail on the Forth Bridges project likewise enabled us to digitally document internationally significant industrial heri-tage in high resolution at a landscape scale.[4]

At the other end of the scale, the HES Collections Team look after over 41,000 col-lections items, a growing number as new items are accessioned or loaned. Variation in dimensions, weight, material and condition also pose a significant challenge for digital documentation. Table 1.2 sets out the range of data capture techniques that we use and approximate tolerances.

Landscape Scale Digital Documentation

The sites and monuments that we digitally document are situated within wider nat-ural and built landscapes that are important to contextualise to better understand their spatial relationships. Landscape-scale 3D data can be used directly to sup-port conservation and condition monitoring projects and enable research where highly detailed topographical information helps to identify patterns such as climate change–related erosion.

We use established remote sensing techniques for landscape-scale data capture. The appropriate method is decided with consideration of the size of the landscape area rela-tive to the required resolution of the data and other factors such as vegetation cover.

TABLE 1.2

Table Adapted from Historic Environment Scotland Short Guide 13 'Applied Digital Documentation in the Historic Environment'

Scale	Specific Technique	Illustrative Accuracy	Output Data Type
Landscape [>km]	Airborne LiDAR	30 mm	Spatial (XYZ), intensity, photo (RGB), orthophoto, classified
	Mobile laser scanning (e.g., boat/vehicle mounted system)	20 mm	Spatial (XYZ), intensity, photo (RGB)
	Aerial photogrammetry (e.g., small unmanned aerial system (SUAS), aircraft)	1–30 mm*	Spatial (XYZ), photo (RGB)
Structure [<km]	Total/Multi-Station	2 mm	Spatial (XYZ), photo (RGB)
	Global Navigation Satellite System (GNSS)	1–5 mm	Spatial (WGS84)
	Terrestrial laser scanning (TLS)	3 mm	Spatial (XYZ), intensity, photo (RGB)
	Structure from motion (SfM) Photogrammetry	3 mm*	Spatial (XYZ), photo (RGB)
	Structured light scanning	0.1 mm	Spatial (XYZ), photo (RGB)
Object [<m]	Triangulation laser scanning	0.05 mm	Spatial (XYZ), photo (RGB)
	Structure from motion (SfM) Photogrammetry	0.1–2 mm*	Spatial (XYZ), photo (RGB)
	Reflectance transformation imaging (RTI)	*n/a*	Photo (RGB), normal map

* Results depend on variables including subject, quality of input images, camera specification and Ground Sample Distance.

Airborne LiDAR

Fixed- or rotary-wing aircraft with onboard laser scanning hardware and image capture can record large swathes of the environment at high altitude. Due to the specialist skills and associated costs, this work is typically commissioned to a commercial contractor. A number of HES Properties in Care have been captured via airborne LiDAR through the Rae Project, at resolutions between 50 and 25 points per m^2. For sites that are linked, such as the properties within the archaeological landscape of Kilmartin Glen, the LiDAR covers both the monuments and the surrounding terrain.

SUAS Photogrammetry

Small unmanned aircraft systems (SUAS) are typically equipped with a digital camera capable of recording high-resolution still images or video sequences. High-spec systems used with a robust flight plan can be capable of achieving very high levels of coverage and spatial resolution for a variety of natural landscapes, including large rock faces.

At Holyrood Park and Edinburgh Castle, we commissioned SUAS photogrammetric surveys to digitally document the natural rock faces at the sites as a baseline for future condition monitoring. This aerial data supplemented terrestrial laser scanning with a total station and GNSS-based control survey. Both data sets are comprised

of approximately 10K full-frame mirrorless camera images, achieving a spatial surface resolution between 1–5 mm with detailed texture maps. Figure 1.3 shows several
views of the Edinburgh Castle rock (and upstanding walls) data set. The high resolution allows inspection of masonry and geological features, including their condition, in a way that would otherwise not be possible or require costly or riskier access
solutions.

FIGURE 1.3   (A) Perspective view of a textured 3D model for Edinburgh Castle surrounding rock and upstanding walls. (B) Mid-range up view of data set showing surface detail.
(C) Close-up detail of the 3D model masonry wall surface.

Site and Monument Scale Digital Documentation

The Rae Project sets out our remit to digitally document all 336 properties under the care of HES. These sites vary significantly in size, complexity, chronology and location across Scotland. This variation amongst our estate can be classified into six categories: (A) roofed monuments that are occupied or in use; (B) roofed monuments that are unoccupied; (C) unroofed monuments with high masonry above 1.5 m; (D) unroofed monuments with low masonry below 1.5 m; (E) standing stones and carved stones; and (F) field monuments (HES, 2015). Our approach to digitally documenting these sites remains flexible but with the objective of capturing robust, survey specification 3D data to support their conservation. We have also worked with a number of external partners to digitally document important cultural heritage sites with similar objectives, and this, likewise, includes subsequent use of the data for engagement, interpretation and accessibility. One example of this is our recent collaborative work with the National Trust for Scotland on The Hill House in Helensburgh, designed in the 20th century as a private residence by architect Charles Rennie Mackintosh. We continue to develop our methodologies with the aim of allowing us to capture more information for these sites (including in some cases additional types of data, such as thermographic) with greater coverage, resolution and accuracy.

Survey Control

To provide a stable frame of reference for Rae Project surveys, permanent survey markers (PSMs) are installed on properties and recorded within the control survey. This supports condition monitoring work across our estate and facilitates the alignment of future surveys. This process is carried out primarily by HES' Building Conservation Technologists with Scheduled Monument Consent in place.

Large (and even 'medium scale') sites can invite compound error due to their size. Sequential alignment of terrestrial laser scans and photogrammetry images throughout a site should be supported with the use of a control survey with the aim of reducing and managing overall error. The control survey should be conducted to within tolerances expected from the overall data set and incorporate all PSMs installed at the property. We primarily use two systems to construct a control survey:

- Total Station—survey instrument with high angular accuracy, enabling point-to-point survey over great distances
- Global Navigation Satellite System (GNSS)—using satellite constellations, including GPS and GLONASS, to provide geolocated British National Grid (OSGB36) coordinates for PSMs

Coordinate values can be processed to accommodate an Ordnance Survey scale factor and depend on the distance to the central meridian. For large sites, this may lead to discrepancy between the GNSS values for control points and their 'local grid' coordinates acquired via a total station. The scale factor used must be explicitly noted in project metadata to ensure appropriate reuse.

Terrestrial Laser Scanning and Photogrammetry

Terrestrial laser scanning (TLS) provides the ability to digitally document interior and exterior environments to a high degree of 3D accuracy and resolution. Systems vary significantly in specification, including speed, accuracy, range and the quality of on-board imaging. We factor in all of these attributes when deciding which scanning systems may be suitable for a specific role, and the following considerations:

- Size and portability, particularly for use in confined spaces or at height
- Integration with control survey via targets and control points
- Levelling, stability and compensation of the system
- Workflow requirements at the data processing stages

As part of the Rae Project, in May 2019 we digitally documented Fort George, a vast 18th-century coastal military fort (Figure 1.4). The site covers 42 acres and includes significant defensive features including outworks, ditches and bastions accessible from the fort's ramparts. Fort George continues to serve as an active military garrison, home to the Black Watch, 3rd Battalion The Royal Regiment of Scotland. To capture the full site exterior and a number of interior spaces, TLS was used extensively as the primary data capture technique, in tandem with a control survey. Over 900 scans were carried out, with a total of over 26 billion points for the fort's exterior spaces point cloud.

Ground and aerial photogrammetry were also captured for selected buildings including the Regimental Chapel (the leftmost free-standing building pictured in Figure 1.4). The Regimental Chapel 3D model is shown in Figure 1.5 with rendered lighting. Considerations for this combined scanning and photogrammetry data capture approach include:

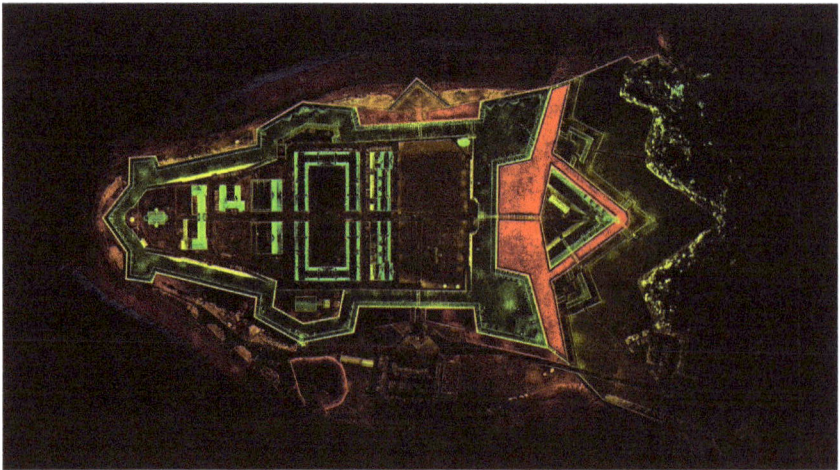

FIGURE 1.4 Fort George terrestrial laser scan data viewed from above in an orthographic projection.

FIGURE 1.5 3D model of the Regimental Chapel at Fort George captured via TLS, ground and aerial photogrammetry during survey.

- Weather conditions and time-of-day leading to lighting differences between laser scans and photogrammetry
- High and unpredictable winds causing potentially difficult flying conditions, as a coastal site surrounded by flat terrain
- Flight permissions, approval and safe access (particularly as an active military site)

Object and Architectural Detail(s) Scale Digital Documentation

Most of HES' collections objects and architectural details (such as carved stones associated with properties) typically have a much smaller envelope for digital documentation. These can range in size from several meters for large carved stones and metalwork, to several millimetres for coins and pins. The required level of detail to properly document these items necessitates high tolerances for 3D capture, well within sub-millimetre resolution and accuracy.

The number of items and their complexity in terms of material properties (which may often be highly reflective, transparent or translucent) can make collections items challenging to digitally document. Care is taken in the handling of objects, using gloves and conservation specification materials to support the objects. To establish a chain of custody of the items within the organisation, an object transfer form records the specific items transferred for 3D capture with direct liaison with the Collections Team. Objects that are located on-site or 'in situ' at a Property in Care

are noted ahead of a Rae Project survey and digitally documented with the appropriate methodology.

Photogrammetry

The flexibility of photogrammetry as a technique allows the adaptation of existing camera equipment for close-range capture. One of the key advantages of photogrammetry is its accurate and extremely high-quality image texture mapping relative to the 3D geometry produced. The choice of lens and capture technique is tailored to the scale of the item. Control of the photographic environment allows the use of motorised solutions to rotate the object and manipulate the camera position, triggering the camera and managing the image pipeline. Figure 1.6 shows an example of a 3D model of a collections item we captured with a 'turntable' style photogrammetry setup. The system can be tweaked to improve data capture of smaller items to a greater level of consistency and accuracy, with the following considerations:

- Macro lenses, extension tubes or reverse lens mounts can enable a lower focus distance to subject and be used with the focus-stacking technique to allow reliable capture of macro-scale items
- Lighting should be controlled, either using 'hot lights' (continuous) with high CRI (colour rendering index) or strobe type lights with diffusers. For sensitive items, lux values should be monitored as light exposure may affect their condition
- Colour-calibration of the camera and lighting configuration will improve the colour accuracy of photography and, therefore, 3D model textures
- Calibrated scale bars with coded targets improve the scaling accuracy and speed

FIGURE 1.6 A photogrammetric 3D model of a prisoner's jewellery box from Dumbarton Castle. Note the glazing covering the fine quilled paper panels. Item accession number 'DUM146'.

- For 'turntable' capture, masking of the input images is often essential to prevent alignment errors arising from the difference in movement between the tabletop and backdrop

Structured Light Scanning

Capable of generating measurable 3D data quickly in real time, structured-light scanning can also accurately record objects in challenging conditions such as in low-light levels or for smooth flat surfaces. Calibrated systems that use real-time tracking to facilitate data capture allow us to quickly capture data in difficult circumstances, such as limited physical access, confined spaces or working at height.

- Accuracy tolerances of the equipment should be known prior to capture, including target resolution and swathe size
- During any post-processing, filtering and final geometry generation stages should be aware of the upper limits of the capture hardware or risk introducing characteristic noise as patterned interference
- The presence of strong ambient light can interfere with white light and infrared systems, reducing their performance during capture

In 2019, HES assisted with the conservation works for the bronze cast of William Wallace at The National Wallace Monument in Stirling (Figure 1.7). The cast stands 6 m in height from its supporting corbel to the tip of the sword and is situated on the southwest corner of the building within a niche, 14 m above the ground level. Due to the immediate need for an accurate, measurable data set to inform the conservation strategy, we opted to use structured light scanning to digitally document the bronze. The initial data capture of the bronze was conducted at height with the use of a mobile elevated work platform (MEWP) and later with scaffold access, using an Artec EVA and Artec Leo 3D scanner. The initial results were used to aid inspection, calculate estimated weight values and ultimately help inform the engineering solution used to remove the bronze from the monument for conservation.

Data Processing and Management

To facilitate the processing of raw scanner or image data, we use a range of open-source and proprietary software. Typically, these are unique to the scanning system or manufacturer and are designed to interpret and process the raw files, which may be proprietary formats with unique file structures. Interoperability is often limited and varied between software packages, in many cases requiring a 'pipeline' across several software packages to fully register and output a completed data set. Software-specific workflows are used to prepare the data, with the use of various algorithms designed to process the data ready for alignment and further use. A general overview of this workflow is shown in Diagram 1.2. Some scanning systems require the complete raw-to-deliverable workflow to be completed in their native software, whilst others can allow export after pre-processing to compatible file formats.

To ensure consistent results across a range of projects at the sites and monuments scale, we combine processed scans into a central software package to produce a database with filtered, clean laser scan data from a range of systems and with support to bring in photogrammetric and SLAM data. During the registration stage, the scan data is aligned using either control (from scanned targets and control survey sources,

FIGURE 1.7   Structured-light scan data of the National Wallace Monument bronze cast.

such as total station and GNSS) and/or 'feature'- based alignment. This uses planar and other geometric features from the point cloud, manually selected 'pick points' and/or visual placement to guide alignment algorithms. The use of the control survey further allows precise georeferencing of the site-scale data to the British National Grid (OSGB36), and accurately control for error at sites of significant scale.

Alignment of 3D scan data with photogrammetric images can be achieved using a growing number of software packages. We use RealityCapture to bring together pre-registered terrestrial laser scanning data sets with aerial and ground-based photogrammetry images. This ensures that the registration characteristics (geolocation

- 'Pre-processing' and filtering (e.g. noise removal, level compensation, colourisation, masking)

- Alignment/registration (and scale transformation if necessary)

- Quality control and refinement

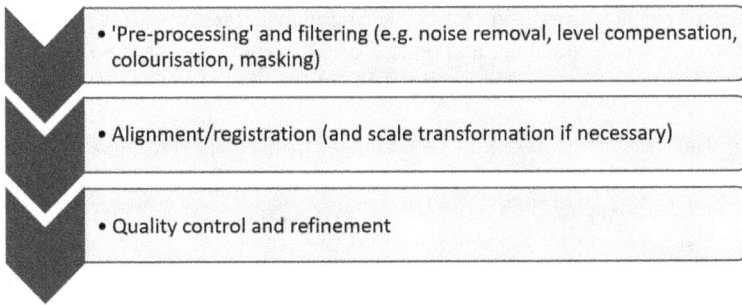

DIAGRAM 1.2   General software-independent workflow stages.

and scan to scan alignment) is maintained as a foundation to align the photography. An essential requirement in this scenario is strong overlap between the two data sets, typically under similar lighting and weather conditions. Some key practices include:

- Inspection of alignment and reconstructing small sectioned areas to pre-view mesh quality
- Use of manually or automatically identified control points to improve any alignment issues between the images and the laser scanning
- Quality control including review of alignment statistics and the mesh and texture quality; artefacts such as 'ghosting' or a 'doubling-up' in the texturing and stepping within the 3D geometry can indicate misalignment
- Selection of best available data; where poor photogrammetry images or redundant scans may be excluded if detrimental to the overall data set or better alternatives exist
- Reconstruction of mesh within the tolerances of the laser scanning and photogrammetry capture

Digital Preservation and Archival Considerations

Ensuring future access to the digital documentation data is essential for its continued use and reuse. As a digital record, the data should remain healthy and accessible to continue to be a valuable asset, enabling research and condition monitoring over extended periods of time. Maintaining stable and secure data are common goals for the digital preservation of information, and specific guidance and advice is published by the UK-based Digital Preservation Coalition.[5]

We advocate general data management principles including:

- The use of open, non-proprietary formats where possible
- Avoiding poorly supported compression algorithms
- Migration of databases to updated stable versions of core software (such as Leica Geosystems Cyclone)
- Storage and organisation of all raw data, project files and outputs generated during the digital documentation project

- Intuitively structured directories and file management
- Retain original metadata and generate summary metadata reports (including field notes and supporting documents)

Backup of all data is an essential requirement to mitigate the impact of data loss from a wide range of possible sources. This can include file corruption from physical damage to the storage medium and encryption from malware. It is equally important to practice version control for new iterations to source data or working data sets to ensure that any working projects do not overwrite raw or original data or otherwise erase hard work.

HOW WE USE DIGITAL DOCUMENTATION DATA

This section will look more closely at some of the digital documentation projects with applied outputs undertaken by the Digital Documentation and Innovation Teams at HES. Our work and the data we produce supports a range of diverse topics, including condition monitoring, interpretation for sites and collections and education and engagement with cultural heritage. The following examples illustrate where we have used our 3D data in recent projects or collaboratively with external partners to facilitate research or promote access to our sites using digital technologies. For further examples of our case studies, please see our 'Short Guide: Applied Digital Documentation in the Historic Environment' (Frost, 2018).

CONSERVATION RESEARCH

We take a collaborative approach to the use of digital documentation data to support scientific inspection and non-destructive materials analysis. From our strong scientific foundations, we continue to work closely with our Conservation Science Team on numerous projects, in the field and in the production of project and research outputs. This includes specific conservation work at some properties in care, such as 17th century Skelmorlie Aisle monument in Largs, and other examples of built heritage including Rosslyn Chapel in Roslin (Wilson et al., 2012) and The Hill House in Helensburgh.

Moisture measurement analysis research recently conducted in Argyle Tower at Edinburgh Castle made use of digital documentation data to visualise the results (Orr et al., 2019). The research aimed to address the potential variation between several moisture measurement techniques, usually caused by the presence of substances such as salts, voids or metals. The approach used an innovative 'data fusion' methodology, combining several data sets into a single index that would factor in the technique characteristics to improve interpretation of the results. The work would also try to identify the presence and possibly the source of moisture ingress within the tower vaulted roof structure.

An excerpt of the results of the work can be seen in Figure 1.8, which shows visual plots of the moisture levels with a simple gradient map. The gradient plots represent the 'homogeneity indices' from three combined techniques: Protimeter electrical resistance, Protimeter capacitance and microwave moisture meter. It is also

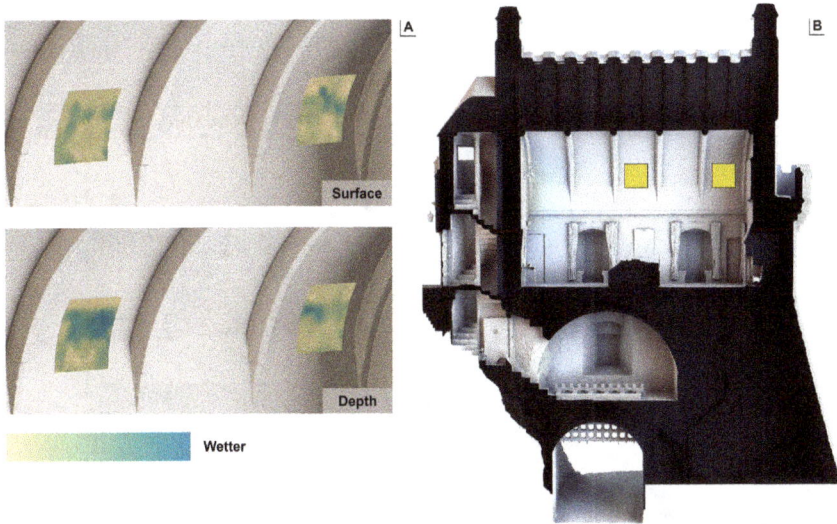

FIGURE 1.8 (A) Moisture at 'surface' and 'depth' level indices overlaid on 3D data of Argyle Tower at Edinburgh Castle. (B) Section style cutaway of Argyle Tower showing the location of the sampling.

Source: Image adapted from Orr et al. (2019)

established that using and interpreting the results requires considerations, such as the mode of construction and materials of the built element and also weather conditions affecting paths of ingress prior to inspection.

CLIMATE CHANGE AND COASTAL EROSION MONITORING

With a wide geographic spread across Scotland, many of our properties in care are located adjacent to or near coastal areas. HES' Climate Change Risk Assessment (HES, 2018) highlights that approximately 10% of our properties are exposed to coastal flooding and erosion in a way that is deemed unacceptable. Approximately 7% of the sites record a 'very high' risk of coastal erosion. Our digital documentation work supports the Scottish Government initiated Dynamic Coast project to establish an evidence base of national coastal change.[6]

Skara Brae is an important settlement that constitutes part of the UNESCO Heart of Neolithic Orkney World Heritage Site. The Neolithic village was initially occupied between 3500–2100 BCE and was later partially exposed in the mid-19th century AD through a destructive storm event. It is located on the Bay of Skaill on the west coast of mainland Orkney. Its location puts it at risk from a variety of environmental factors including storminess and sea level rise, both directly related to ongoing climate change. A sea wall defence was erected in the early 20th century (and later strengthened) to mitigate erosion of the site. However, this does not protect the adjacent dune areas, which are unprotected and exposed and may contain unexcavated archaeology.

To help monitor coastal erosion at the site, the HES Digital Documentation Team has undertaken biennial terrestrial laser scanning with specific focus on the coastal

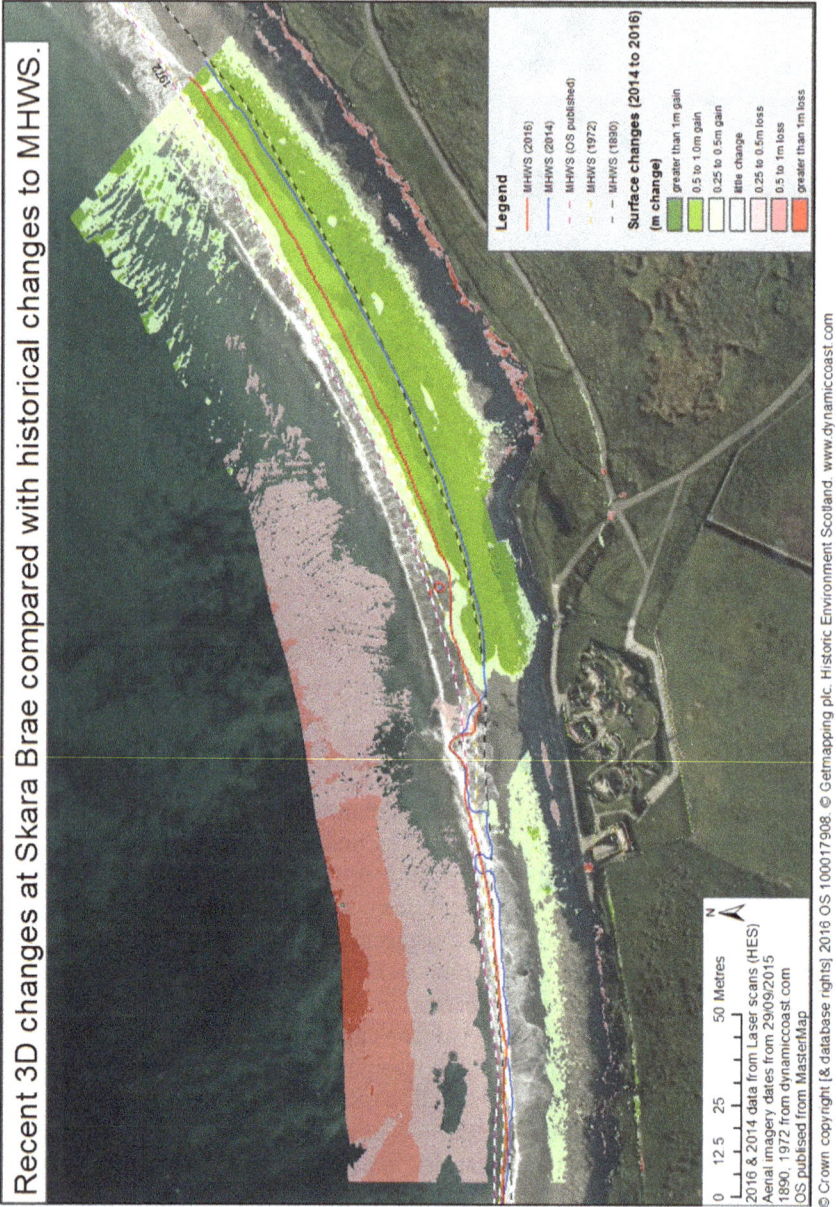

FIGURE 1.9 Recent 3D changes at Skara Brae compared with historical changes to MHWS.

Source: © Crown Copyright

area. Scanning epochs in 2010, 2012, 2014, 2016 and 2018 have produced data sets that act as a baseline for deviation mapping to indicate material gain and loss within this area. Figure 1.9 shows an illustration produced by the Dynamic Coast project team using data HES TLS gathered between 2014 and 2016. The gain and loss of material is shown relative to the mean high water spring (MHWS), showing the changing sea level relative to the coast. As part of the Rae Project, this commitment will continue to produce data to aid the monitoring of the coastal area and serve as a crucial baseline for identifying changes and trends to the site's local environment.

ACCESSIBILITY

A practical use of digital documentation data is to facilitate virtual access to sites in our care. At Maeshowe Chambered Cairn, part of the UNESCO Heart of Neolithic Orkney World Heritage Site, entry to the small 5,000-year-old chamber adorned with elaborate Viking runes is via a low 14.5 m long passageway. For some visitors, this is not physically possible. Additionally, only 20 people can fit snugly inside the chamber on a guided tour. The popularity of the site means we often cannot accommodate visitor demand. To address this, we have developed a virtual access app (*Explore Maeshowe*) based on laser scan and photogrammetry data captured as part of the Scottish Ten project. The 3D data was post-processed to produce a photo-textured model and then dropped into Unity game engine. The result is an explorable environment where visitors can have a virtual experience of the interior and exterior of the cairn, finding out about the site and its place in the rich Neolithic landscape of Orkney (Figure 1.10). The app is freely downloadable on iOS and Google Play stores, and at the time of writing has had 5,820 downloads. To accompany the app, we also have a version available on VR headsets in the Maeshowe visitor centre, which have been used by our visitor services team to give completely virtual guided tours. While certainly no replacement for visiting the real site, for those who are not able to physically access for whatever reason, the app and VR experience provide a viable alternative, for which we have received very positive feedback.

SUMMARY

At HES, the fundamental driver for our work is how we make digital data useful and meaningful for the understanding, protection and enjoyment of our historic environment. The use of the digital technologies discussed in this chapter is central to how we deliver digital projects, data and guidance in alignment with Scottish Government and HES Corporate priorities. Strategic thinking guides this approach and is supported by our ability to demonstrate strong practical applications for these digital technologies within the historic environment. The use and reuse of these digital documentation data sets enable a diverse range of uses within and outside the organisation and illustrate the value for money invested in the process.

Conducting and facilitating research and innovation have remained the core purpose around which we deliver projects. It drives our exploration and adoption of new technologies in the field and the studio but also helps us tread new ground and address questions that arise as part of the process. However, a number of challenges

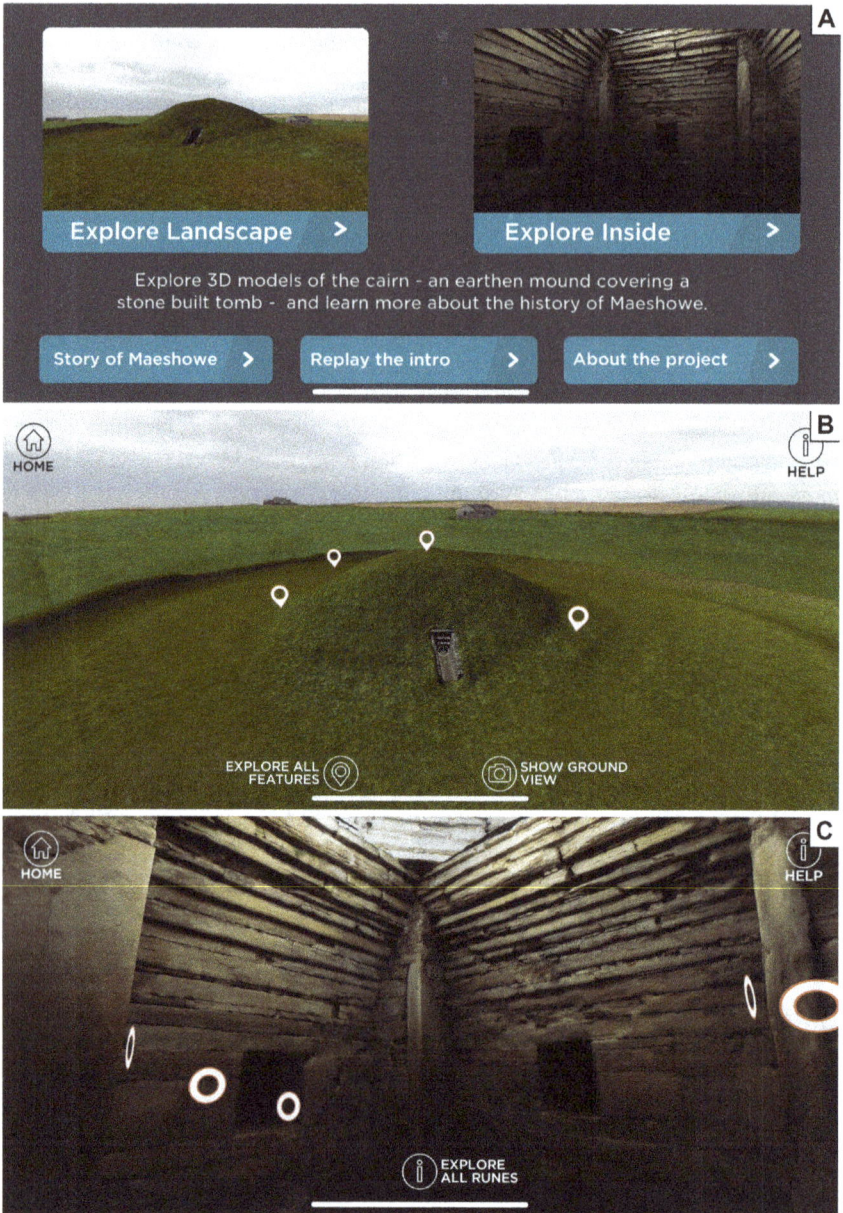

FIGURE 1.10 *Explore Maeshowe* app, using 3D digital documentation data in a gaming environment.

remain for us, including upskilling those connected with the discipline, which will help to support our long-term vision and promote relevant skills and knowledge within the wider sector. Through our various engagement channels including publications, collaborative partnerships, traineeships, academic support and our public programme at The Engine Shed, we are able to widely deliver advice and guidance. This is integral to our overarching goal of mainstreaming and normalising the use of digital technologies within the heritage sector. We continue to work towards this ambition and see the growth and adoption of these digital technologies for the benefit of the historic environment.

ACKNOWLEDGEMENTS

The work of the Digital Documentation and Digital Innovation Teams at HES would not be possible without the efforts of all of our team members. These are Adam Frost, James Hepher, Sophia Mirashrafi, Al Rawlinson, Alan Simpson and Dr. Lyn Wilson. We are grateful for the assistance and support we receive from our colleagues across the organisation, many of whom we have been able to work with in the field as well as the office.

We would also like to thank our external partners, some of whom have been featured in the case studies in this chapter. This includes the National Trust for Scotland, Scottish Natural Heritage, the National Wallace Monument, University College London Institute for Sustainable Heritage, The University of Edinburgh, Heriot Watt University, and The Glasgow School of Art.

NOTES

1. www.engineshed.scot/about-us/the-scottish-ten/
2. www.engineshed.scot
3. https://github.com/CyberbuildLab/masonry-cc
4. Scottish Government, 2017 and www.theforthbridges.org/visit/go-forth-digital-learning-resources/
5. www.dpconline.org/
6. www.dynamiccoast.com

REFERENCES

Frost, A. (2018) *Short Guide 13: Applied Digital Documentation in the Historic Environment.* Historic Environment Scotland: Edinburgh. Available online: <www.historicenvironment. scot/archives-and-research/publications/publication/?publicationId=9b35b799-4221-46fa-80d6-a8a8009d802d>

Hepher, J., Wilson, L. and Antonopoulou, A. (2016) The Rae Project: Digital Documentation of a Nation's Heritage. In May, K., editor, *Digital Archaeological Heritage: Proceedings of the International Conference, Brighton UK, 17–19 March 2016. EAC Occasional Paper No. 12,* 19–23. Europae Archaeologia Consilium, Belgium.

Historic Environment Scotland (2014) *Our Place in Time. The Historic Environment Strategy for Scotland.* Scottish Government. Available online: <www.historicenvironment.scot/ archives-and-research/publications/publication/?publicationId=fa088e13-8781-4fd6-9ad2-a7af00f14e30>

Historic Environment Scotland (2015) *Baseline Condition of the Properties in the Care of Scottish Ministers*. Available online: <www.historicenvironment.scot/media/4626/hes-baseline-condition.pdf>

Historic Environment Scotland (2018) *Climate Change Risk Assessment*. Available online: <www.historicenvironment.scot/archives-and-research/publications/publication/?publ icationId=55d8dde6-3b68-444e-b6f2-a866011d129a>

Historic Environment Scotland (2019a) *Digital Strategy 2020–23*. Available online: <www.historicenvironment.scot/archives-and-research/publications/publication/?publicationi d=eb3e5276-3887-4b03-8fbb-ab2200bcd8c2>

Historic Environment Scotland (2019b) *Heritage for All: Corporate Plan 2019 Onwards*. Published by Historic Environment Scotland. Available online: <www.historicenvironment. scot/archives-and-research/publications/publication/?publicationId=1f65f457-a602-4ddc-af61-aa2500933d61>

Historic Environment Scotland (2019c) *Research Strategy*. Published by Historic Environment Scotland. Available online: <www.historicenvironment.scot/archives-and-research/publications/publication/?publicationid=ea6de1b6-f5db-4b1a-bfab-aa1e01036c1e>

Historic Environment Scotland (2020) *Climate Action Plan 2020–25*. Published by Historic Environment Scotland. Available Online: <www.historicenvironment.scot/archives-and-research/publications/publication/?publicationId=94dd22c9-5d32-4e91-9a46-ab6600b6c1dd>

Orr, S., Young, M. and Frost, A. (2019) Using Multi-Sensor Moisture Measurement in Conservation through 'Building Pathology Indices' and 3D Digital Documentation. *Monuments in Monument Conference*, Stirling, 31 January. Available online: <https://pub-prod-sdk.azurewebsites.net/api/file/e99b4f10-3572-4459-a81f-aabc00ab0e4e>

UNESCO (2019) *UK National Commission for UNESCO*. ISBN: 978-0-904608-07-6. Available online: <www.unesco.org.uk/wp-content/uploads/2019/07/Cultural-Heritage-Innovation-2.pdf>

Valero, E., Bosché, F. and Forster, A. (2018) Automatic Segmentation of 3D Point Clouds of Rubble Masonry Walls, and its Application to Building Surveying, Repair and Maintenance. *Automation in Construction*, **96**, pp. 29–39.

Valero, E., Bosché, F., Forster, A., Hyslop, E., Wilson, L. and Turmel, A. (2019) Automated Defect Detection and Classification in Ashlar Masonry Walls using Machine Learning. *Automation in Construction*, **106**, p. 102846.

Wilson, L., Pritchard, D.K., McGregor, H.C. and Mitchell, D.S. (2012) Two Avenues for Data: Rosslyn Chapel as a Terrestrial Scanning Case Study. Proceedings of the XXII Nordic Surveyors Congress, Oslo 2012. *Kart og Plan*, **72**, pp. 315–320.

Wilson, L., Rawlinson, A., Mitchell, D.S., McGregor, H.C. and Parsons, R. (2013) The Scottish Ten Project: Collaborative Heritage Documentation. *The International Archives of the Photogrammetry, Remote Sensing and Spatial Information Sciences*, **XL-5/W2**, pp. 685–690. https://doi.org/10.5194/isprsarchives-XL-5-W2-685-2013.

2 Mapping the Urban Environment with a Handheld Mobile LiDAR System—A Case Study from the UrbanARK Project

Aaron Miller, John Meneely, Ulrich Ofterdinger, Debra Laefer, Michela Bertolotto, and Anh Vu Vo

THE URBANARK PROJECT

This research project is a tripartite collaboration between Queen's University Belfast, University College Dublin and New York University. Its goal is to enhance flood risk management for urban coastal communities using LiDAR applications. Worldwide, floods are a major threat, with widespread social, economic and environmental impacts. Today, over 600 million people live in critical coastal zones, and nearly 66% of the world's cities with more than 5 million inhabitants fall within such areas (McGranahan et al., 2007). This risk is increasing annually, particularly in coastal areas due to accelerating mean sea-level rises, combined with increasing population growth coupled with new construction in flood-susceptible areas.

Urban coastal flood inundation models are typically based upon coarse digital elevation models/digital terrain models (DEM/DTM) generated from airborne LiDAR data. These airborne data sets vary in spatial resolution and, by the nature of their collection, give little or no information on underground areas or building facades, missing important details which can be indicative of underground structures and their dimensions. As a result of this low density of information, many urban features are not captured, such as street furniture, kerbs or minor variations in surface elevation.

Previous investigations into the level of detail in these 3D models have illustrated that small-scale features have a significant impact on flood propagation and surface water flooding in urban environments. Surface water inundation is often more rapid when finer resolution models are used due to the rapid propagation of water along 'channels' that form at the road edge as a result of the road camber and roadside kerbs (Fewtrell et al., 2011).

This study will employ handheld mobile laser scanning (HMLS) technology to map part of the Central Business District (CBD) in Belfast, N. Ireland (Figure 2.1) at a considerably higher resolution than existing airborne LiDAR, which was last collected in 2006 at a spatial resolution of 0.5 m. Aerial LiDAR covers large areas very

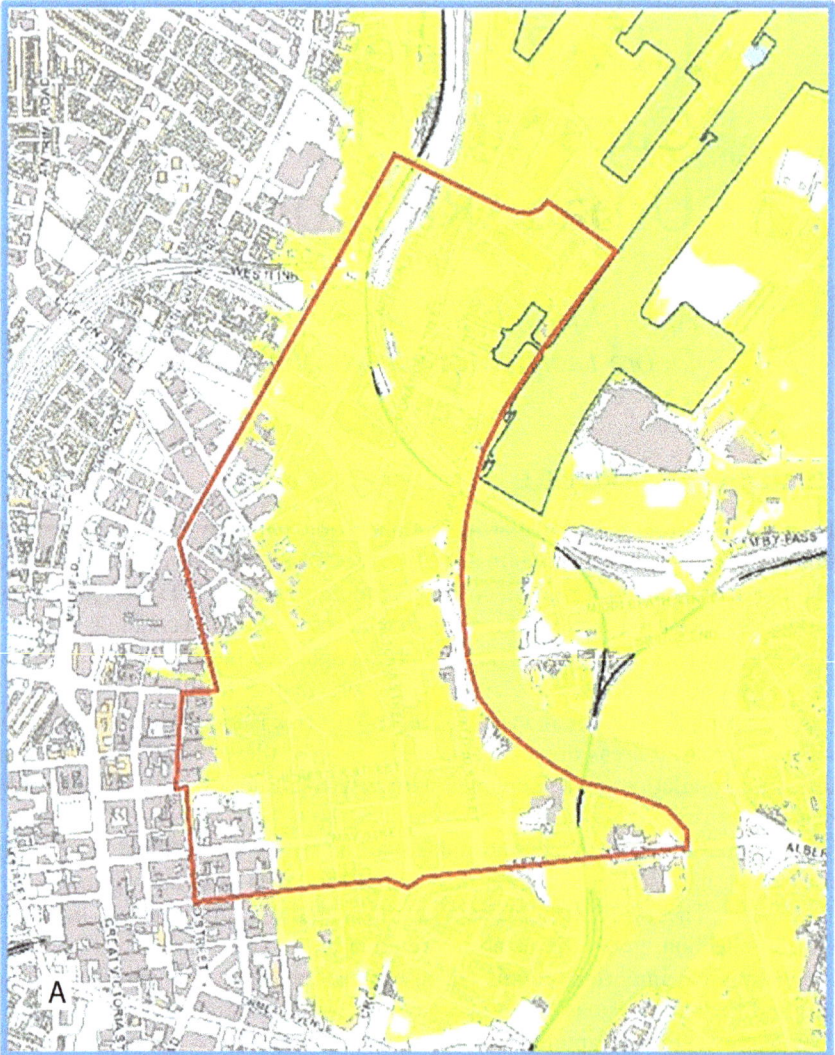

FIGURE 2.1 (A) Study area (red boundary). The light green region is the floodplain predicted under future climate change scenarios. This is based on a 3D model generated from airborne LiDAR data collected in 2006 and represents 1-in-200-year coastal flood events (EU flood directive report, 2007). (B) Areas within the study area, highlighted in yellow, with underground structures extracted from planning applications for the period 1973–2018.

(*Continued*)

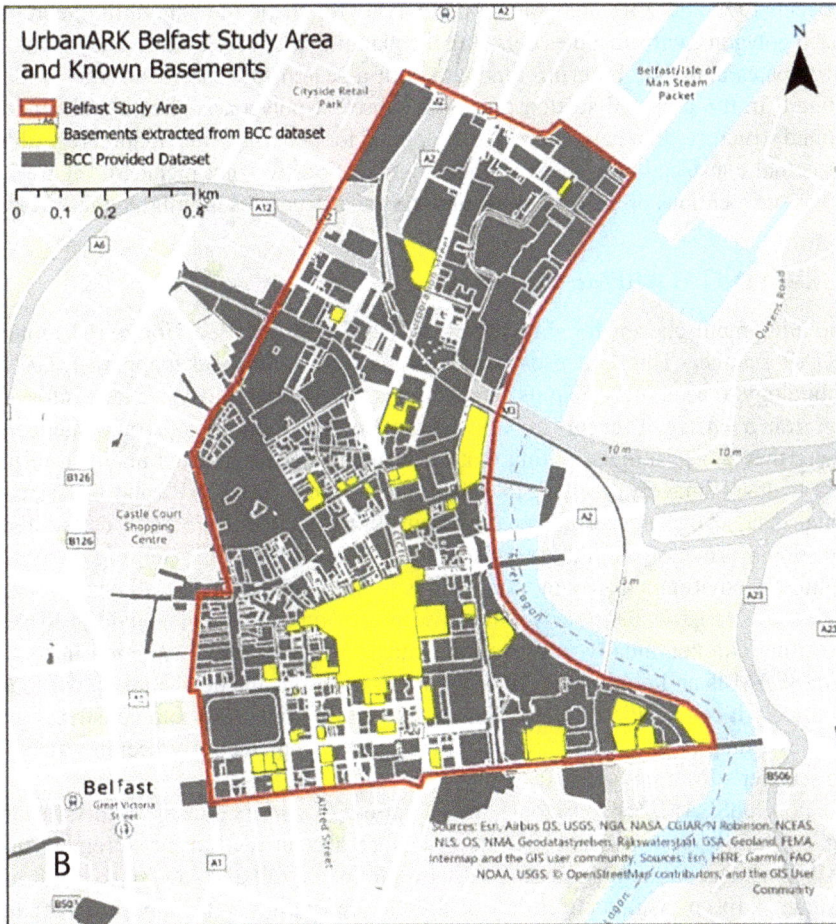

FIGURE 2.1   (Continued)

quickly, but it is very expensive. The 3D flood prediction base model produced from this HMLS data for the study area will be more detailed, less costly (both financially and environmentally), up-to-date and contain information on the location and geometry of some underground structures. Other advantages of adopting this technique include the ability to quickly re-survey areas of new construction, keeping the model current. Also, narrow streets, covered alleyways, pedestrian underpasses and street furniture that is difficult to identify in low-density airborne LiDAR or even vehicle-mounted LiDAR systems can be included in a 3D model as these play a significant role in modelling flood wave propagation and inundation rates.

Prior to any HMLS street-level mapping, a GIS desk-based study was undertaken to identify known underground structures in the study area. The data set used to find these locations was supplied by Belfast City Council (BCC). This information was derived from planning applications provided to BCC by the Department for Infrastructure (DfI). It contained information regarding all planning applications between 1973 and 2018 and was supplied as a GIS shape file containing georeferenced polygons with attributes detailing the planning proposal, status, date, location, coordinates and size. Filtering this data with a search for 'basement' and 'underground' in the proposal section created a subset of polygons indicating all underground structures that have successfully applied for planning. Information regarding the actual construction of these approved applications was not included. Figure 2.1 shows the locations of these possible underground structures highlighted in yellow.

SURVEYING EQUIPMENT

The equipment chosen for this survey is the GeoSLAM Zeb Horizon™ mobile LiDAR scanner. This unit uses simultaneous localisation and mapping (SLAM) technology to accurately map its surroundings. SLAM-based devices take information from a variety of sensors to build a picture of the environment around them and where they are positioned within that environment. These sensors may use visual data (camera imagery), non-visible data (sonar, radar or LiDAR) and basic positional data, using an inertial measurement unit (IMU) or GPS. The device then uses the information from these sensors to calculate a 'best estimate' of where it is within the environment. By moving its position within this space, all environmental features (e.g., walls, floors) will move in relation to the device and the SLAM algorithm can continually improve its estimate with this new positional information. SLAM is an iterative process—the more iterations the device takes, the more accurately it can position itself within that space. In brief—this unit constructs and updates a map of an unknown environment while simultaneously keeping track of the scanner's location within it.

This GeoSLAM Zeb Horizon™ uses a variety of sensors to achieve this, including a rotating, time-of-flight, 3D laser scanner and an inertial measurement unit (IMU), which continually measures its attitude (pitch, roll and yaw) in space. It has a range of 100 m, with an accuracy of 1–3 cm (depending on the environment) and collects 300,000 points per second. Each measurement is stored as an x,y,z coordinate with an intensity value and the time of measurement. Intensity, usually recorded as value between 0–1, is how much of the laser power that was sent out to measure a point returns from that point. This intensity value can be used for material differentiation, but care must be taken when doing this as it also depends on the angle of incidence and the distance to a target. These metrics combined with the mobility of a handheld scanner, where it can effortlessly be taken from street level into a structure or underground spaces such as carparks, stairwells, underpasses and basements while continually capturing data, make it suitable for surveying an urban environment efficiently.

Figure 2.2 is an image of the device. It consists of a relatively lightweight scanner head (1.5 kg), usually held out in front as you walk along, and a data logger (1.3 kg), which is worn separately over the shoulder or in a backpack. It can also be mounted on a backpack, bicycle, car or drone. Colour information can also be mapped onto the scan data during post-processing with the addition of an 'action' type camera placed just below the laser scanning unit or a 360° camera placed on a pole above the system. Colour information was not collected during this study.

FIGURE 2.2   Image of the GeoSLAM Zeb Horizon™ HMLS unit and reference plate.

Using this technology, city blocks can be scanned in a matter of minutes com-pared to a terrestrial, tripod-mounted 3D laser scanner (TLS), which may offer much higher levels of accuracy and point density but could take hours/days to survey the same area. Figure 2.4 shows a section of a street in Belfast scanned with both the HMLS unit and a FARO TLS.

The HMLS unit comes with a reference plate, which attaches to the bottom of the handle and allows the user to collect reference points during a survey. This is achieved by setting the scanner on the ground, upright and stationary for a period of five seconds with the reference plate crosshairs (red circle in Figure 2.2) over the intended point to be recorded. These stationary periods are automatically detected during data post-processing and the time, coordinates and orientation data for each stationary episode are recorded as a reference point. These reference points were accurately surveyed prior to collecting HMLS data with a Leica GNSS survey grade system. When post-processing the scan data in GeoSLAM Hub™, the user needs to enter the coordinates for the detected reference points (either imported from a file or manually) to perform a transformation. It is recommended to mark and measure these reference points with a GNSS system prior to undertaking the HMLS survey, as collecting reference points 'on the fly' may result in these locations being in urban

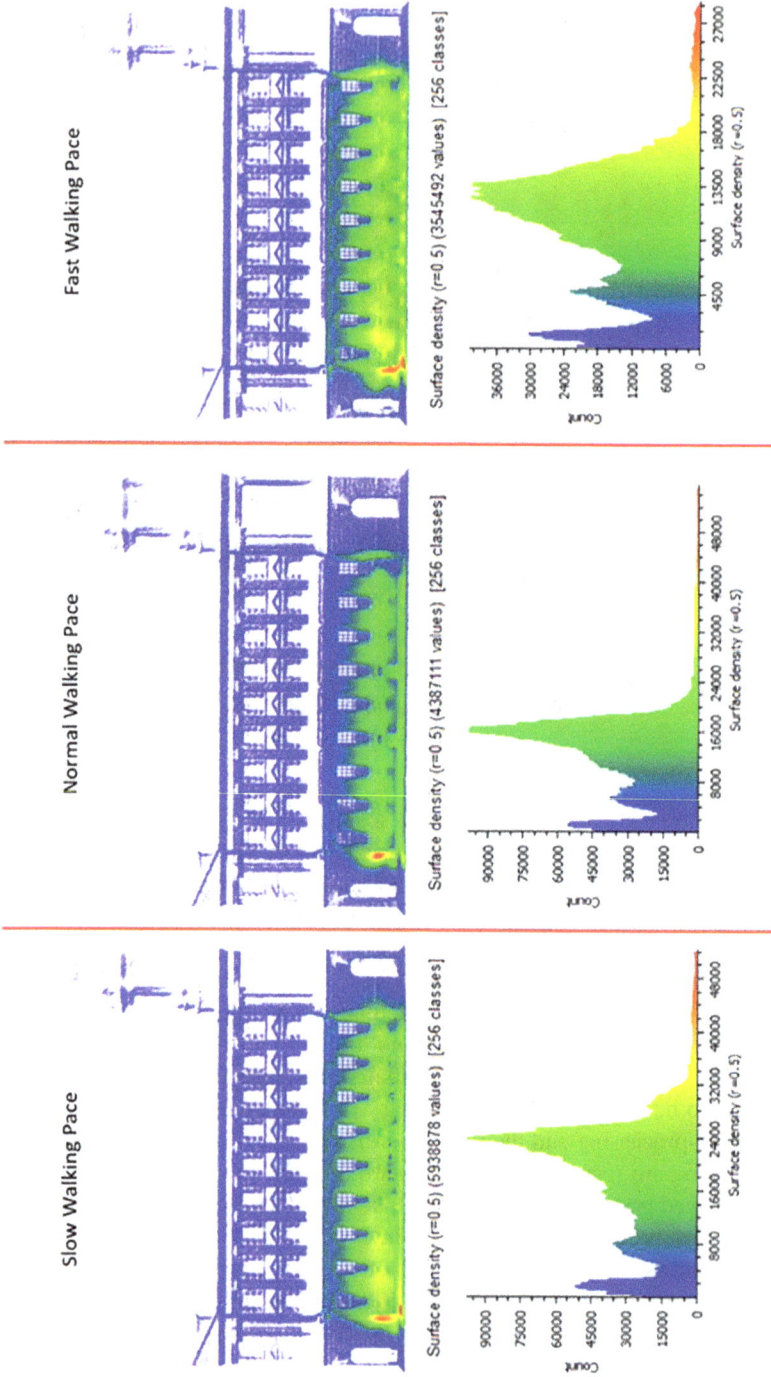

FIGURE 2.3 Trial of data density collected at 3, 4 and 5 kph. Data density was calculated using CloudCompare software by measuring the number of surveyed points within a circle of 0.5 m radius (r) of each point.

canyons—areas of poor GPS coverage at street level due to tall buildings blocking the incoming signal (Gabela et al., 2020).

One drawback of this and other SLAM-based systems is that surveys must be collected in a closed loop (i.e., they must start and finish at the same place). As a result of this, scanning routes should be planned in detail. The data is also collected in its own coordinate system; therefore, it must be converted to a known coordinate system (georeferenced) during post-processing.

To start a scan, the unit is switched on and placed on the ground for approximately 30 seconds allowing the IMU to calibrate. Once this process is complete, the unit is picked up and held steady in one hand in front of the operator while the pre-planned route is walked, stopping to pick up pre-surveyed reference points where necessary, returning the unit to its starting point before switching off. As this investigation takes place in the urban environment, the decision was made to split individual surveys into city blocks, with a maximum survey time of 20–25 minutes—the manufacturers recommend a maximum individual survey time of 30 minutes.

Positioning of the operator on the footpath/sidewalk during a survey has a minimal impact on the quantity of data collected at street level. As the laser scanner measures by line of sight, it is advised, where possible, to walk along the outside edge of the footpath, closest to the road to collect more data on the upper portions of buildings. The main interference in data collection in cities will come from pedestrians, which can come between the operator and the target. Therefore, surveys should be planned for when the area is at its quietest with respect to people. Initial trials were carried out to determine the effect of the operator's walking speed on data density and the accuracy of the data by comparing it with a TLS survey. As this scanner collects measurements at a set rate of 300,000 per second, the density of the data collected is determined by the speed of the scanner through space—slower speed equals denser data. Figure 2.3 shows the data captured on a section of Belfast City Hall at three different operator walking speeds—3, 4 and 5 kph or slow, normal and fast. The building façade is coloured from blue (lower density) to red (higher density) data. The difference in scan density, shown in Table 2.1, for each speed was as expected, with the highest number of points collected in the slow-paced survey, fewer in the 'normal' speed scan, and the least in the fast-paced survey.

TABLE 2.1
Scan Density Results from the Walking Speed Trials

Walking Speed Kph	Slow—3 kph	Normal—4 kph	Fast—5 kph
Number of points on surface (millions)	5.9	4.3	3.5
Peak surface density—pts/ circle with a radius of 0.5 m	25,000	16,000	13,500
Number of pts with peak surface density	92,000	63,000	40,000

Predictably, a fast walking speed collects less data, and a slower walking speed collects more data at a higher density, but the time taken to collect this is longer. Based on this trial, a normal walking pace was selected as it provided an adequate density of data at a flood-relevant height.

To determine the accuracy of the HMLS system, a comparison survey using a tripod-mounted TLS system was carried out on part of Elmwood Ave, Belfast. The avenue was first surveyed with the HMLS by simply walking along one pavement, crossing the road at the end and returning down the opposite pavement to the starting point. It was then surveyed with a FARO Focus™ 3D laser scanner—in total 20, 5-minute colour, static scans were collected on both sides of the avenue. Both sets of LiDAR data were then processed in their respective manufacturer's software—HUB for the GeoSLAM™ and SCENE™ for the FARO data. Registration of the TLS data produced an error of 2 mm across the whole survey. Both data sets were then converted into ReCAP format for import into Autodesk's AutoCAD, where a series of comparison measurements were made (Figures 2.4 and 2.5).

Figure 2.4 shows a block of houses extracted from the two comparison data sets. The top data set (A) is from the TLS, with the measured points coloured with RGB values, and the bottom set (B) is from the HMLS and its data set is coloured by intensity values. This image shows that over the length of this block the HMLS was 20 mm longer. Figure 2.5 is a cross-section of both data sets at the same location, with the TLS data on top and the HMLS below. This image shows a series of comparative measurements across both surveys. The distance between the two streetlights in the TLS data is 21.02 m, in the HMLS data it is 21.01 m. The difference in height between one of the streetlights and the pavement below is 0.01 m, while the slope angle of the roof on the building on the left of the image was $32°$ in both data sets. These results, which fell within the manufacturer's stated tolerances, confirmed that the accuracy of the HMLS system would be more than satisfactory for surveying in the urban environment.

As this system maps using SLAM technology any sudden, jerky movement of the scanner may cause the sensors, especially the IMU, to lose their location in space

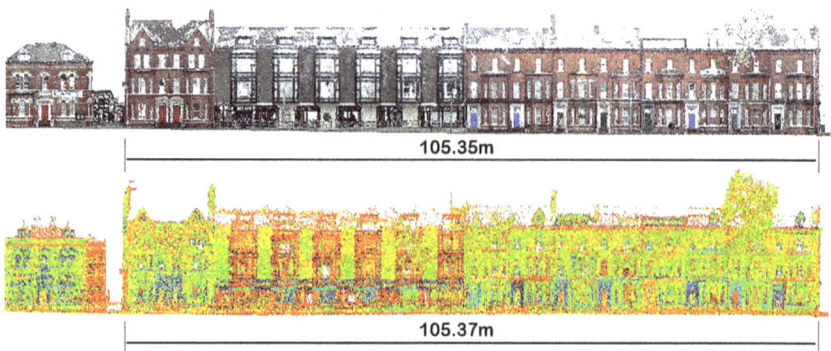

FIGURE 2.4 Comparison of TLS (top) and HMLS (bottom) on a block of houses on Elmwood Ave, Belfast, imported into AutoCAD.

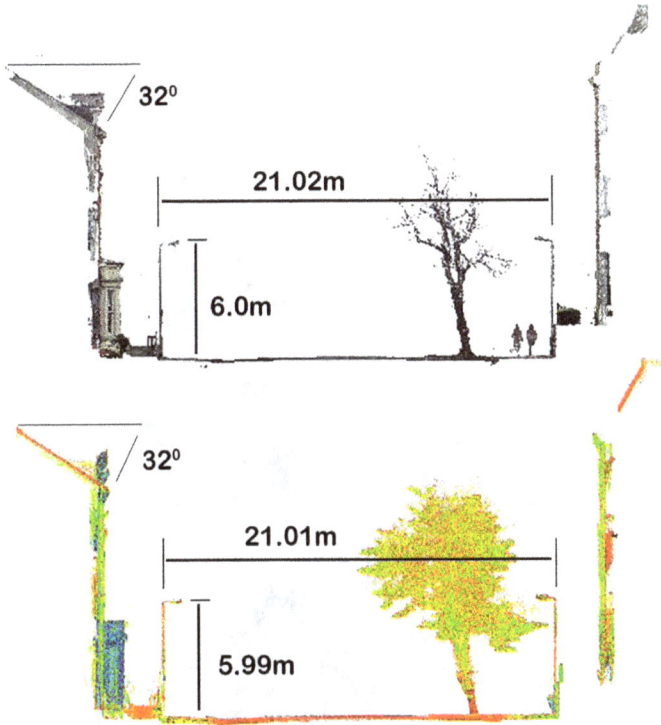

FIGURE 2.5 A cross-section of Elmwood Ave, Belfast, surveyed with a TLS (top) and HMLS (bottom).

and the survey will fail. It is important to reiterate that the success or failure of a survey is unknown until the data has been post-processed in the manufacturer's software. It is recommended that this post-processing be carried out as soon as possible, preferably on-site, in case a repeat survey is needed. However, if the survey fails to process with a good SLAM result, it is possible to change the influence of the sensors used during this processing (e.g., increase the weight given to the IMU versus LiDAR data during processing) to achieve a satisfactory result.

Once the individual data sets are post-processed with a positive SLAM result, it is possible to register (join) up to four of these surveys together using a merge function in the manufacturer's software. This is achieved by approximately placing the scan's locations relative to each other in a graphical interface. The merge function then detects matching corners, edges, planes, etc. between these scans and accurately joins them into a single data set. This merged scan is then georeferenced using the GNSS-surveyed reference points, exported in LAS format, and opened in a GIS software for further analysis. In total, 42 individual scans were collected in the 1 km² study area. Figure 2.6 is a plan view of this resulting data generated in CloudCompare. This figure is colour-coded by height with higher values in red.

FIGURE 2.6   All the data collected with the HMLS in the study area.

DETECTION OF UNDERGROUND SPACES

Prior to surveying with the HMLS, an additional desktop study was undertaken to investigate the efficacy of using the existing low-density airborne LiDAR, collected in

2006, to identify underground structures in the study site. The airborne LiDAR data and a 'bare earth' (buildings removed) digital terrain model (DTM) were imported into GIS software. The DTM was then used as a cutting plane, removing any data above and leaving only points that were below it and, therefore, possibly below ground. These two data sets are available free to download from www.opendatani.gov. uk. This 'first step' in identifying underground spaces from the 2006 airborne LiDAR data had very limited success due to the density and age of the data—as some of the buildings in the study area were constructed post-2006 and, as previously stated, the dynamics of airborne LiDAR collection. However, this information can still be useful in identifying possible underground spaces in areas where the HMLS data could not reach (e.g., in courtyards surrounded by buildings or behind high walls).

Figure 2.7 is a section of the 2006 Belfast LiDAR data and a Google Earth image of the same area. In Figure 2.7 (A) the LiDAR data is colour-coded to height with RGB values and the 'bare earth' DTM is coloured black and opaque.

In the top-down plan view, any points below the DTM and, therefore, below ground are not visible and show up as black areas, see Figure 2.7 (A). The red circles indicate two areas that are possibly underground. The Google Earth image in Figure 2.7 (B) with the same highlighted areas shows that these are entrances to a pedestrian underpass. However, both images only give the entrance/exit location with no indication of the depth or geometry of this subterranean feature. By comparison, Figure 2.8 is HMLS data of the area shown in Figure 2.7. It clearly demonstrates the advantages of surveying in the urban environment with an HMLS over airborne LiDAR in that the pedestrian underpass, identified from the desktop study, could be accessed and accurately mapped, giving its depth below ground, geometry and the ability to calculate its volume.

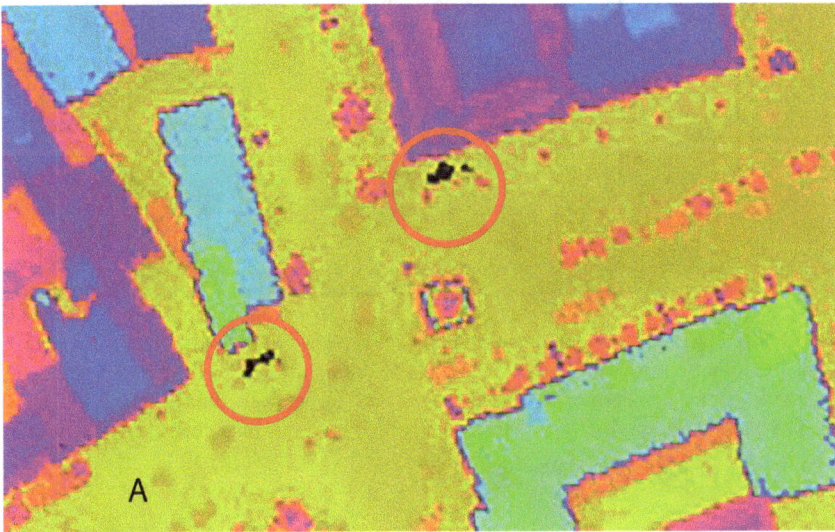

FIGURE 2.7   (A) section of the 2006 Belfast LiDAR data set. (B) Google Earth Pro 2022 Image Landsat/Copernicus of the area in (A).   (*Continued*)

Source: Image generated in CloudCompare

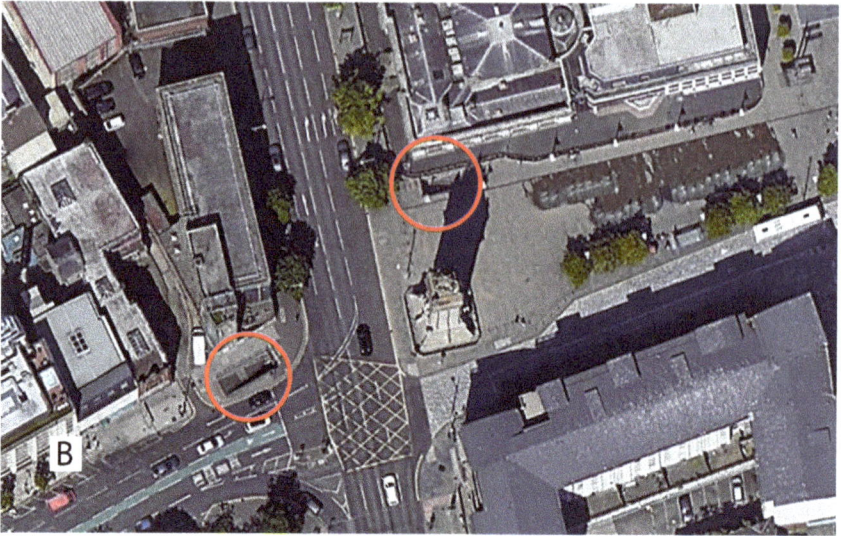

FIGURE 2.7   (Continued)

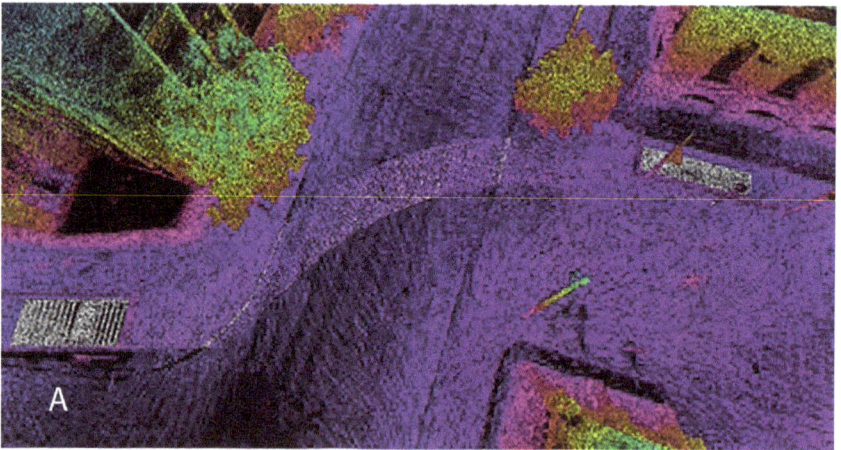

FIGURE 2.8   (A) This shows a top-down perspective view of the HMLS data—above DTM points are colour-coded with height, below DTM points are white. (B) and (C) are ortho-graphic views of the underpass in plan and elevation, respectively. Image (D) shows the 3D model of the underpass with the dimensions of a bounding box in metres.   (*Continued*)

Source: Generated in Microsoft Meshmixer.

B

3.105m

124.076m²

C

23.721m²

2.703m

2.745m

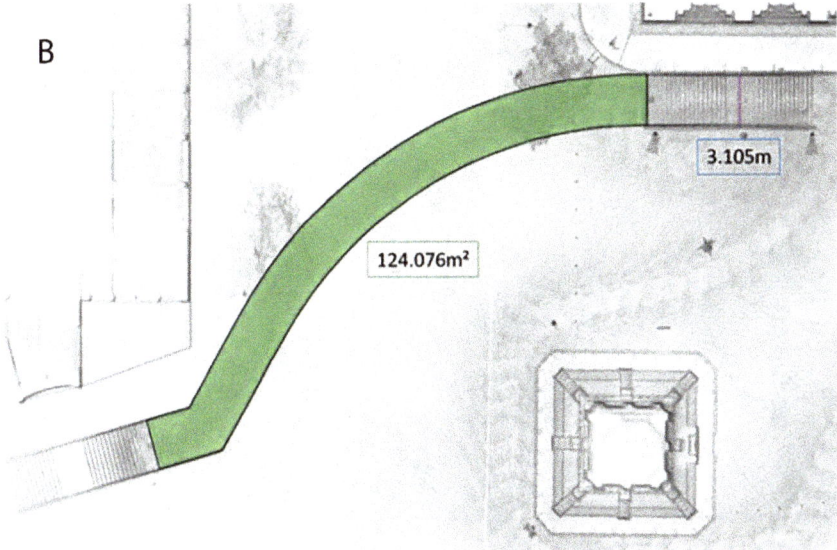

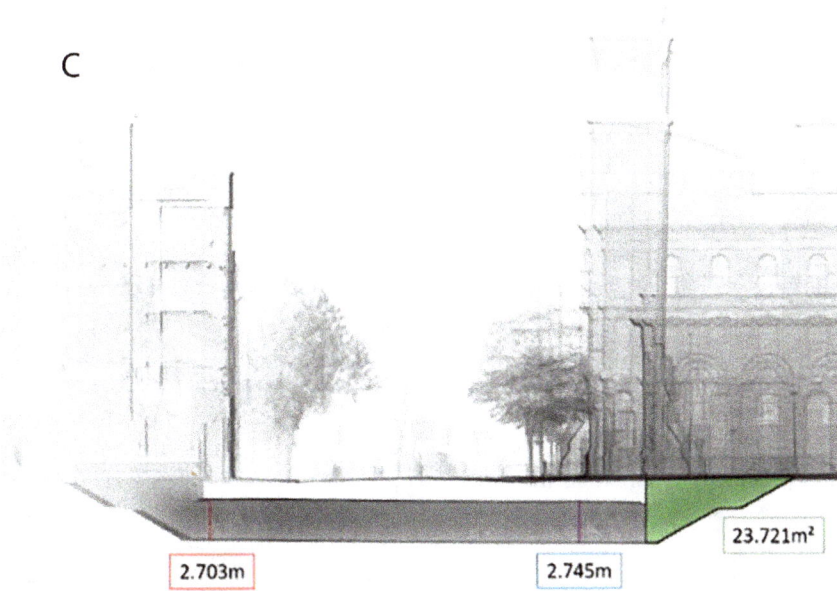

FIGURE 2.8   (Continued)

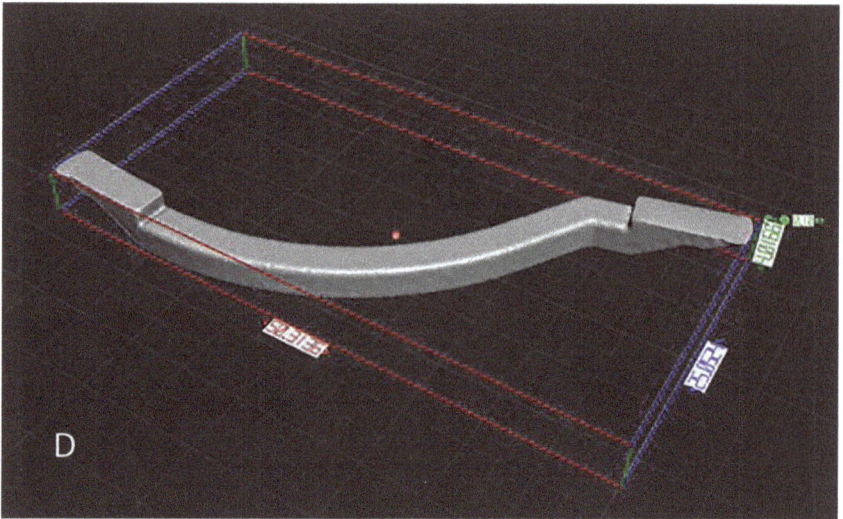

FIGURE 2.8 (Continued)

The volume of this underpass is 446.6 m³. It was calculated by converting the point data of the underpass into a mesh, using a poission surface reconstruction function in CloudCompare and then importing this mesh into Meshlab,[1] which has a volume calculation function. It is important to note that including these small 'sinks' in a coastal flooding model will have no effect on the final water level reached during such an event as the sea is treated as an infinite water supply, but it will certainly influence the rate of flooding through the urban environment. Knowing the internal geometry of these spaces may also allow an estimate of the inundation rate into them and, thus, give an indication to the emergency services as to how rapid evacuations need to proceed.

As previously stated, the resolution of a flood inundation model has a dramatic effect on predicting the speed of flood wave propagation through it. Figure 2.9 (A) is a Google Earth Pro image of an area known locally as the Belfast Entries (highlighted in red in Figure. 2.9). These 'Entries' are a series of historical narrow alleyways in the city centre that run between High Street and Ann Street, and when the town was first laid out, these alleyways serviced dense residential and commercial development. Portions of these entries are less than 1 m wide and most have covered entrances or sections, making it very difficult to survey or identify with airborne and vehicle-mounted LiDAR systems. Figure 2.9 (B) is a 20-minute survey with the HMLS system, and it visibly illustrates the advantages of adopting this technology to survey this type of urban feature over the existing 2006 airborne LiDAR data for this area shown in Figure 2.9 (C). When this higher level of detail on the location and internal geometry of these alleyways is included in a flood inundation model, it will greatly improve its accuracy of how water will flow through the city centre during a coastal flooding event.

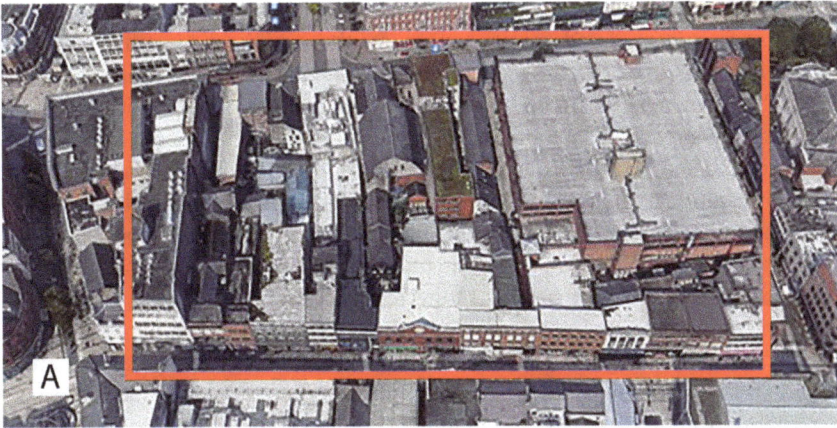

FIGURE 2.9   (A) Google Earth Pro 2022 Image Landsat/Copernicus of Belfast's Entries.

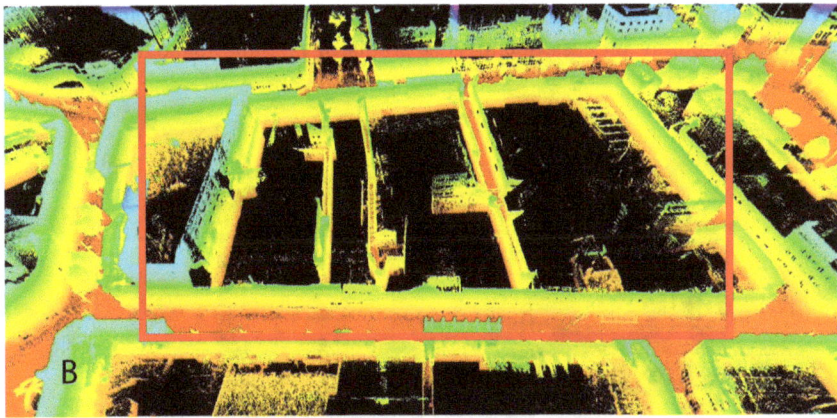

FIGURE 2.9   (B) HMLS survey of Belfast's Entries.

FIGURE 2.9   (C) 2006 airborne LiDAR of Belfast's Entries.

Not all underground spaces could be entered and surveyed with this technique. Some private owners of underground structures that have been identified during the desktop investigation or in the survey data have declined a request to survey their space. Therefore, a method based on extracting as much information on the location and possible geometry from the 3D scan survey has been developed.

Figure 2.10 shows three examples of the types of evidence that will be used for basement identification and measurement where access was not possible. On closer investigation, Building A has no survey data below ground, but it does show windows at street level, indicating a possible underground structure. Not included in this case study but part of UrbanARK's future research plan is to investigate the use of feature recognition software to automatically identify and locate these street-level features. As no below-ground data has been gathered on this building, an estimate of the height and extent of a basement will have to be made from the footprint of the building. Buildings B and C in Figure 2.10 are examples of where underground spaces were identified by HMLS data below the 'bare earth' DTM discussed previously. Building B shows a basement level that is accessible via a stairway from the street. The stairway, windows, doorway and basement level were all captured during the HMLS scanning process, identifying points of possible inundation. This data

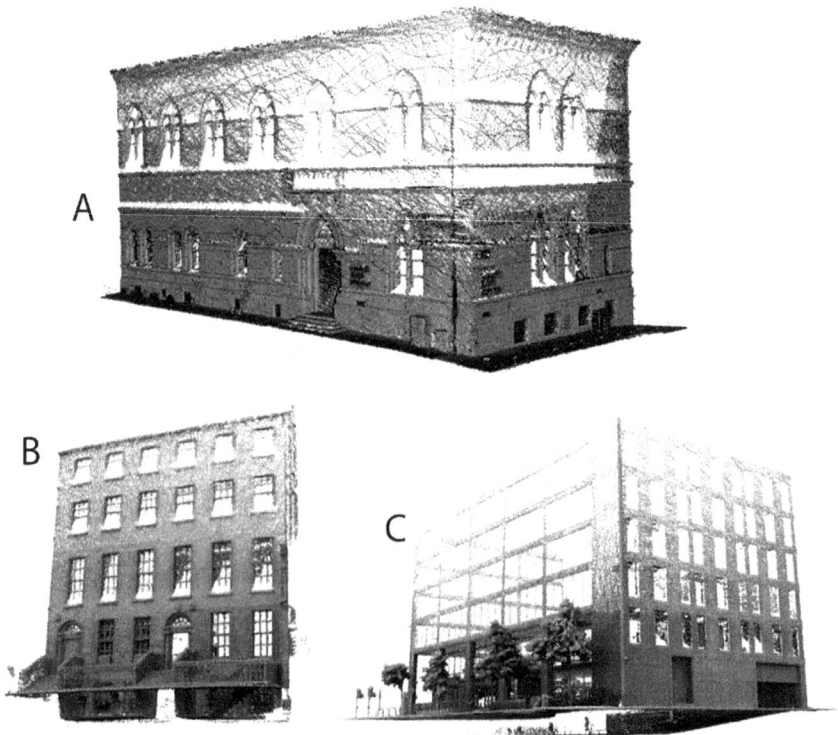

FIGURE 2.10   HMLS data was collected on three buildings in the city with pavement—DTM ground level (red line).

will give the depth of the basement under this building but not its extent. Building C shows the entrance to an underground parking structure, which was open at the time of surveying and allowed for data to be collected down an entrance ramp, to its base, giving its depth and some information on its extent.

Recent advances in 'high density, low-level' aerial LiDAR may allow the identification of the features identified in Buildings A and B, but not C. An example of this type of data collected by Laefer et al. (2017) over a 2 km² area of Dublin CBD from an altitude of 300 m, with a point density of 335/m² can be viewed at www.youtube. com/watch?v=qEi2Wo7Bcuk. But, even this exceptional data set comes with the aforementioned caveats of cost, environmental impact, the geometry of underground structures and keeping it up-to-date.

The identification of these possible points of inundation during a flooding event will play an important role in modelling the flow of water through the city. It is planned to give these locations a 'sink value', and if/when the wavefront of a modelled flood event intercepts one of these points, an algorithm can remove a volume of water until the void it has encountered is filled.

CONCLUSION

HMLS has proven to be a cost-effective method for mapping the urban environment for the creation of 3D flood models, with a much higher point density at street level than airborne LiDAR; detailed information of building facades, which may indicate the presence of basements; and greater ease of use when surveying underground structures. The extent to which this HMLS urban data set will impact future flood simulations is yet to be seen, but as a result of improved terrain modelling, the inclusion of narrow alleyways, street furniture, underground structures and points of inundation into them, these simulations can only be more accurate and improve the risk awareness of residents and emergency services.

ACKNOWLEDGEMENTS

Funding for this project was provided by the Northern Ireland Department for the Economy (DfE) as part of the project 'UrbanARK: Assessment, Risk Management, & Knowledge for Coastal Flood Risk Management in Urban Areas' (USI 137), jointly funded under the US-Ireland Research and Development Programme with the US National Science Foundation (NSF Award 1826134) and the Science Foundation Ireland (SFI—17/US/3450).

NOTE

1. www.meshlab.net

REFERENCES

Fewtrell, T.J., Duncan, A., Sampson, C.C., Neal, J.C. and Bates, P.D. (2011), Benchmarking urban flood models of varying complexity and scale using high resolution terrestrial LiDAR data. *Physics and Chemistry of the Earth, Parts A/B/C, 36*(7–8), pp. 281–291.

Gabela, J., Kealy, A., Bachelet, X., Moran, W. and Hedley, M. (2020), Evaluation of integrity availability based on classic RAIM in different urban environments for stand-alone GPS and multi-sensor solutions. *Conference Paper: International Global Navigation Satellite Systems Association IGNSS Symposium 2020*, Sydney, NSW, Australia.

Laefer, D.F., Abuwarda, S., Vo, A.V., Truong-Hong, L. and Gharibi, H. (2017), *2015 Aerial Laser and Photogrammetry Datasets for Dublin, Ireland's City Center [Data set]*. New York University. Center for Urban Science and Progress. https://doi.org/10.17609/n8mq0n.

McGranahan, G., Balk, D. and Anderson, B. (2007), The rising tide: Assessing the risks of climate change and human settlements in low-elevation coastal zones. *Environment and Urbanization*, *19*(1), pp. 17–37.

3 Using Drones to Map and Visualise Glacial Landscapes

Iestyn D. Barr, Kathryn Adamson, Timothy Lane, Konstantin Nebel, and Willem G. M. van der Bilt

INTRODUCTION

As glaciers retreat from the landscape, they leave behind a suite of characteristic landforms (e.g., moraines, eskers, flutes, terraces, and fans) (Figure 3.1), which can be mapped and analysed to reconstruct past glacier dimensions and dynamics (Pearce et al., 2017; Chandler et al., 2018). This approach has a strong heritage that extends over 150 years but has rapidly evolved over recent decades in response to the availability of high-resolution remotely sensed data (e.g., satellite imagery and digital elevation models). Such data are particularly useful in glacial environments since they allow large and often remote areas to be rapidly and systematically mapped by a single person (or group). The most recent development in this field is the use of unmanned aerial vehicles (UAVs), which are now routinely adopted for glacial geomorphological mapping (Chandler et al., 2016; Tonkin et al., 2016; Ely et al., 2017; Le Heron et al., 2019). The key advantage of UAVs is that they allow high-resolution (< 0.1 m per pixel) imagery to be captured in a relatively quick, flexible, and inexpensive way, including allowing repeat (sub-annual to annual) surveys to be undertaken (Immerzeel et al., 2014; Chandler et al., 2016). Their use has been particularly useful for investigating recently deglaciated landscapes (i.e., those that have been exposed from beneath glaciers within recent years-to-decades), because features in such landscapes are often subtle (i.e., small) and short-lived (i.e., they may only last a few years), since they are susceptive to considerable erosion/modification as the landscape rapidly responds to deglaciation.

Studies using UAVs to map glacial landscapes typically fall into one of three groups: (1) studies focused on reconstructing glacier dimensions (e.g., Chandler et al., 2016); (2) studies investigating landform morphometrics (size and shape) (e.g., Ely et al., 2017); and (3) studies looking at changes in land surface elevation over time (e.g., Tonkin et al., 2016). Despite these differences, the overall purpose of UAV surveys in glacial environments is typically to generate high-resolution orthophotos and digital elevation models (DEMs). As a result, the approach to data acquisition and processing is similar across studies.

DOI: 10.1201/9780429327575-3

FIGURE 3.1   Characteristic glacial landforms (including moraines, eskers, flutes, fans, and terraces) in the forefield of Hørbyebreen Svalbard.

In this chapter, a general approach (framework) for collecting, processing, and analyzing UAV data in glacial environments is outlined before a sample of case studies are described.

FRAMEWORK FOR DATA COLLECTION, PROCESSING, AND ANALYSIS

A number of studies have developed operational frameworks for acquiring and processing UAV data (Westoby et al., 2012), some of which relate specifically to glacial environments (e.g., Ewertowski et al., 2019; Lamsters et al., 2019). Here, we build on these overviews and highlight a number of key steps to successfully collect, process, and analyse UAV data in glacial landscapes.

UAV SELECTION

The choice of UAV for a particular survey is often determined by the overall study purpose and the data sets required as outputs (e.g., optical vs. thermal imagery) (Tang and Shao, 2015; Harvey et al., 2016). However, since glacial environments are often remote and inaccessible (e.g., high mountain, high latitude), ease of transport is also a key consideration when selecting an appropriate UAV. For this reason, small, lightweight UAVs, such as the DJI Phantom and Mavic series (Figure 3.2), have been particularly popular for studies of glacial landscapes (e.g., Lamsters et al., 2019; Storrar et al., 2019). These UAVs are not only easy to transport in the

FIGURE 3.2   DJI Mavic Pro (circled) immediately prior to a UAV survey at Hørbyebreen Svalbard.

field (e.g., in a rucksack) but also benefit from having batteries of a size and power capacity (< 100 Wh) that allow them to be carried as hand luggage with most airlines.

SITE AND UAV PREPARATION

Prior to beginning a UAV mission, it is necessary to contact local authorities, land-owners, and nearby airports. Though this is less important in glacial environments, which are often remote, it is still advisable to provide precise information about fly-ing location and flight times, particularly when there is the potential for low-flying helicopters (a risk in mountainous areas).

Once in the field, the steps to site preparation vary depending on the overall pur-pose of the study. If the purpose is to generate a glacial geomorphological map from which glacier dimensions and dynamics might be inferred (Chandler et al., 2016), then precise (e.g., cm accuracy) locational information isn't necessarily needed, and the accuracy of the satellite positioning system onboard the UAV (typically accuracy ~ 5–10 m) will often suffice (Turner et al., 2013; Ewertowski et al., 2019). However, if the study purpose is to extract metrics from small glacial landforms (e.g., flutes—Ely et al., 2017) or investigate change detection between multiple surveys (Tonkin et al., 2016), this requires that data sets (orthophotos and DEMs) are under-pinned by highly accurate and precise locational information. To achieve this, one of the first steps is to collect ground control points (GCPs), which serve as tie-points for resulting orthophotos and DEMs (Tonkin and Midgley, 2016).

When GCPs are required, where possible, they should be evenly distributed over the study area, ensuring that the full altitudinal range is covered. The number of GCPs required for a particular survey varies according to site size and characteristics (e.g., topographic variability), however, it has been suggested that each GCP has a functional radius of ~100 m (Tonkin and Midgley, 2016). For each GCP, the precise location should be measured in the field using a differential global position system. GCPs are often based on purpose-made targets, with distinct centre-points. However, as an alternative, or an addition, notable features in the landscape (e.g., distinctive boulders) can function as GCPs but often lack distinct centre-points, thereby introducing greater errors during processing (Tonkin and Midgley, 2016; Ewertowski et al., 2019). Once the location of each GCP has been recorded in the field, approximately two-thirds of this data set are ideally used to georeference the data, while the remaining third is used as independent checking points to assess overall error (Ewertowski et al., 2019).

To conduct missions, UAVs can be flown manually or by pre-programming flight paths (e.g., grids) using freely available mobile applications (e.g., DJI GO, Drone Harmony, or Pix4Dcapture) (Lamsters et al., 2019). Benefits of pre-programming include being able to specify flight altitude (with apps automatically calculating the resulting ground sampling distance) and being able to specify the desired overlap between images. Whether flying manually or following a pre-programmed route, a number of simple equipment checks should be performed prior to takeoff. In particular, checking that propellers are correctly mounted and undamaged; checking that sensors are operating correctly (if not, calibration may be required); and checking that the return-to-home altitude and home-point location are set.

FLYING A MISSION

Most glacial environments lack dense and/or tall vegetation, hence flights are often conducted within visual line of sight (VLOS). This is facilitated by a two-person flying team, with one person flying the UAV and one acting as a 'spotter'. Since the purpose of most glacial landform studies is to produce an orthophoto and DEM, vertical (nadir) images are sufficient, though high image overlap is required (often > 70–80%). This requirement for high overlap means that thousands of photos are typically captured during a single field season/visit.

Once flying a mission, it is advisable to return the UAV when the battery level drops below a pre-determined setting. In glacial environments, it is important to be conservative and start to return UAVs with comparatively high battery levels (e.g., 30%—Ewertowski et al., 2019). This precaution is suggested since glacial environments are often characterized by strong and gusty winds (which can quickly drain battery life), and by rugged ground, where forced/emergency landings (e.g., due to insufficient battery life) are particularly risky. In an attempt to reduce the chances of running out of battery life, a few key steps are often followed. First, when flying a pre-determined grid (e.g., in Pix4DCapture), the mission is often started at the furthest position from the home-point location (i.e., the grid is flown towards the operator). This means that at some point during the flight, the operator can easily judge whether wind conditions are likely to prevent the grid from being fully completed and can then abort the mission and return the UAV. Second, where possible, it is advisable to fly-out against the wind, and return with it (i.e., set a grid so that the furthest point, where the mission starts, is up-wind from the home-point). This will reduce the likelihood of running out of battery power as the UAV fights a strong headwind on its return home.

Issues relating to battery power are particularly relevant when surveying glacial environments since access to charging facilities is often limited. Glacial environments are also cold, which can limit battery performance. Precipitation, combined with strong and variable winds (many UAVs are unable to fly in high wind speeds) means that conditions in glacial environments are rarely conducive to easy flying, and in many cases, flying time is restricted by weather conditions and/or battery life (either the UAV or controller batteries) (Ely et al., 2017).

In glacial environments, UAV takeoff and landing can be particularly challenging due to widespread and unstable fields of boulders (one reason to avoid emergency landings, whenever possible). To address this, purpose designed, foldable landing pads are often useful since they help avoid rocks and prevent last-minute indecision about where to land. In some cases, 'hand' takeoff and catching can also be performed (whereby someone holds the drone before takeoff, and catches it, mid-air, afterwards), though this is restricted to UAVs with attached landing gear (e.g., the DJI Phantom series) (e.g., Lamsters et al., 2019), and should be performed with extreme caution.

DATA PROCESSING

Irrespective of the software used, processing UAV-derived photographs from glacial environments typically follows standard procedures. Here, these procedures are illustrated using Agisoft Metashape (formerly Agisoft Photoscan) photogrammetry software. If a laptop with suitable software is available, it is often advisable to run Steps 1 and 2 in the field (e.g., after returning to a field-station/camp in the evening), this allows data (image) gaps to be identified and targeted during subsequent missions.

Step 1: Blurred, overexposed (often a problem over glaciers), or otherwise 'unsuitable' photos should be deleted from the data set (this is where image overlap becomes important in providing a large sample of photos from which a number might need to be deleted). Most photogrammetry software allows image quality to be determined automatically (Sieberth et al., 2016), and the user can then select a quality threshold suitable to their needs (e.g., removing all images with a quality below 0.7). **Step 2:** Once unsuitable images have been removed, the remaining photographs are 'aligned' (based on common features in overlapping images). This generates a sparse point cloud (Figure 3.3 (A) (B)), which can be edited to remove outlier points manually and/or using automated methods—e.g., points can be selected and deleted on the basis of their reprojection error, reconstruction uncertainty (Figure 3.3 (A)), image count, and/or projection accuracy. **Step 3:** Locational (x, y, z) information can be incorporated from GCPs (if available/required). **Step 4:** From the edited sparse point cloud, a dense point cloud (Figure 3.3 (C), Figure 3.4) can be generated, which should again be edited manually and/or using automated

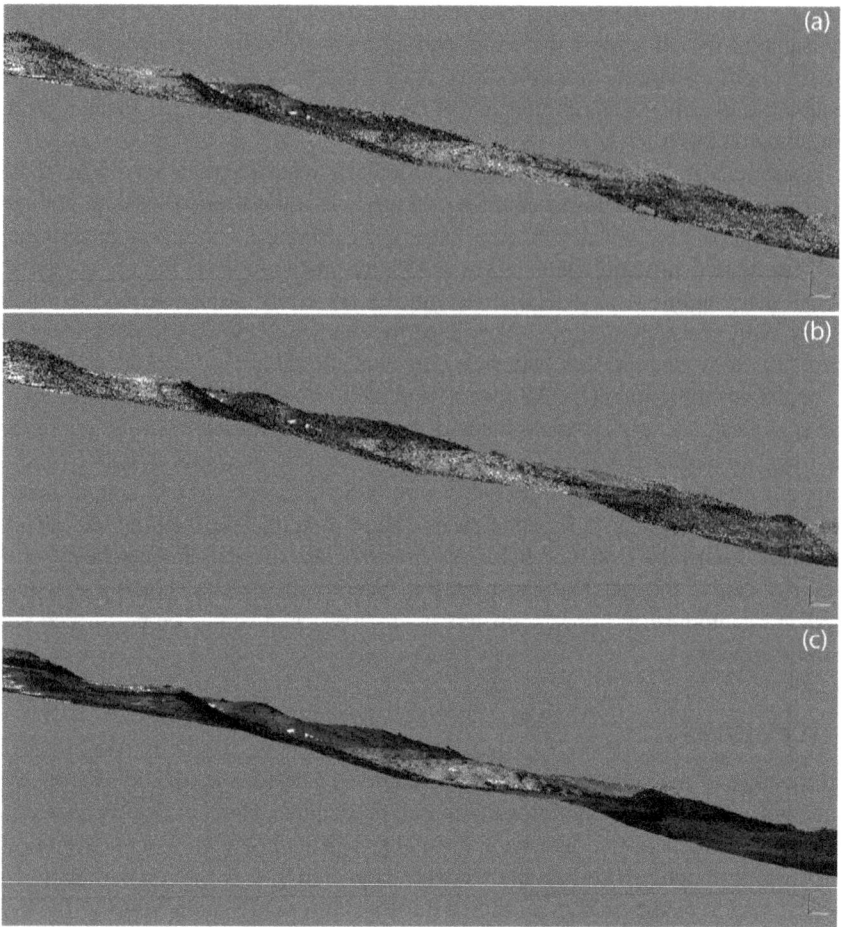

FIGURE 3.3 Point clouds of the foreland of Vatnsdalsjökull (Iceland). (A) Sparse point cloud where points with a high reconstruction uncertainty are highlighted in pink. (B) Sparse point cloud following deletion of highlighted points from 'A'. (C) Dense point cloud generated from the sparse point cloud in 'B'.

methods. **Step 5:** From this dense point cloud, a DEM and orthophoto can be created (Figure 3.5) and exported in a user's preferred format and projection.

DATA ANALYSIS

Once an orthophoto and DEM have been generated, they can then be analysed in numerous ways, and the steps involved vary according to the overall purpose of a study. For studies purely focused on glacial geomorphological mapping (for the purpose of glacier reconstruction), there are standard procedures for mapping from both orthophotos and particularly DEMs (see Chandler et al., 2018). These procedures typically include on-screen identification and digitisation of different glacial

FIGURE 3.4 Oblique view of a dense point cloud of the foreland of Vatnsdalsjökull (Iceland).

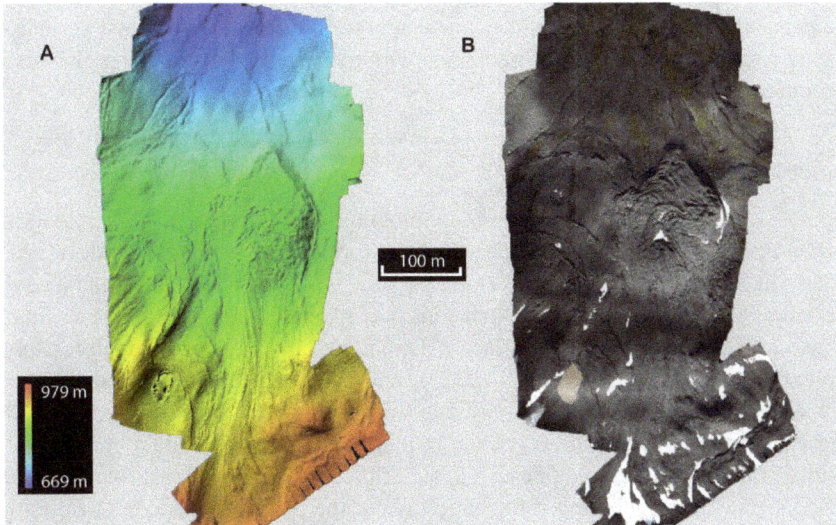

FIGURE 3.5 (A) DEM and (B) orthophoto of the foreland of Vatnsdalsjökull (Iceland), generated from the dense point cloud in Figure 3.4.

landforms, including moraines (Chandler et al., 2016), eskers (Storrar et al., 2019), drumlins (Allaart et al., 2018) flutes (Ely et al., 2017), and other features (Le Heron et al., 2019). For studies focused on measuring and analyzing landform morphometries, precise mapping is required prior to statistical analysis of landform size and shape (e.g., Ely et al., 2017). This requirement for precision is why GCPs are vital is such cases. For studies using multiple DEMs from different time periods to look at changes in surface elevation over time (Tonkin et al., 2016; Storrar et al., 2019), it is vital that DEM co-registration is performed and that uncertainty in resulting outputs is quantified (Noh and Howat, 2015).

CASE STUDIES

Here, some published studies that demonstrate the utility of UAVs in glacial environments are briefly described. The first example, Chandler et al. (2016), is a study focused on reconstructing glacier dimensions. The second, Ely et al. (2017), investigates landform morphometrics (size and shape). The third, Tonkin et al. (2016), investigates changes in land surface elevation over time.

Landform Mapping and Reconstructing Glacier Dimensions

Chandler et al. (2016) used a UAV to map small-scale, recessional push moraines in the foreland of Skálafellsjökull, SE Iceland, to better understand past glacier dimensions and dynamics. The imagery captured has a spatial resolution of 0.09 m, allowing these small moraines that mark former ice margin positions to be readily identified and mapped with a precision impossible from lower resolution data. The resulting moraine maps were used to reconstruct past glacier fluctuations (when combined with chronological information) and demonstrate that glacier fluctuations are particularly sensitive to summer air temperatures as well as that past periods of glacier retreat at Skálafellsjökull coincide with the retreat of other outlet glaciers in Iceland and Greenland.

Landform Morphometrics

Ely et al. (2017) used UAV-derived photographs to generate a high resolution (~0.02 m horizontal resolution) DEM of the foreland of Isfallsglaciären (a small glacier in Swedish Lapland). The study purpose was to identify, map, and analyse the morphometry (length, width, and relief) of glacial landforms known as flutes, which are indicators of former glacier dynamics. Flutes are small features (e.g., typically < 20 cm tall, Ely et al., 2017), and are often rapidly eroded from the landscape (i.e., within decades following their formation). This means that UAVs allow them to be analysed in ways and in numbers not possible via other methods. In total, Ely et al. (2017) mapped 88 glacial flutes and extracted information about their morphology (including information about the presence/absence of initiating boulders at their stoss ends). On the basis of this work, Ely et al. (2017) suggest that examples from Isfallsglaciären generally support the lee-side cavity infill model of flute formation (Boulton, 1976). This approach to analysing landform morphometrics from UAV-derived imagery has considerable potential, which has yet to be fully exploited.

STUDY OF SURFACE ELEVATION CHANGE

Tonkin et al. (2016) compared a UAV-derived DEM from 2014 to a LiDAR-derived DEM from 2003 in order to quantify surface elevation change in the forefield of Austre Lovénbreen, Svalbard, (particularly its ice-cored moraines). During this 11-year period (i.e., 2003–2014), they found an average surface lowering of −1.75 ± 0.89 m, and consider this to reflect the ablation of buried proglacial ice. They found spatial variability in the extent of this lowering and attributed this to differences in the quantity of buried ice within moraines. This study demonstrates the usefulness of UAVs in monitoring and quantifying how glaciers and adjacent areas evolve and contribute to sea-level rise, as the global climate warms.

FUTURE DIRECTIONS

The use of UAVs in glacial landscapes has seen considerable and rapid adoption over recent years, and UAVs have made notable contributions to our understanding of glacial environments (in ways that had previously been impossible). At present, UAVs are largely used for landform mapping, but their use is likely to develop beyond this in coming years. Particular areas of future interest are likely to include the repeated, multi-year surveying of glacial forelands (ultimately with multi-annual data sets covering decades), to monitor and quantify how such landscapes evolve during and following deglaciation (Tonkin et al., 2016; Midgley et al., 2018). There is also potential for making stronger and more direct links between UAV-derived data and data derived from ground-based surveys. This is true of studies which use UAVs to analyse surface conditions and ground-penetrating radar and/or sedimentology to investigate the sub-surface (Storrar et al., 2019). At present, poor weather conditions and short (< 30 min) battery life limit UAV flight times. We can do little about the former, but the latter will no doubt see considerable future improvement.

CONCLUSIONS

The use of UAVs to investigate glacial landscapes has seen considerable development and widespread adoption over recent years. In fact, because UAVs allow small-scale (sub-metre scale) features to be identified and can be repeatedly deployed (e.g., in multi-year surveys), they have fundamentally changed the way that recently deglaciated landscapes (i.e., those in front of modern glaciers) are mapped and monitored. To date, much of this work has involved using UAV-derived data to reconstruct the dimensions and dynamics of former glaciers; to analyse the morphometrics of glacial landforms (to better understand how they form); and to monitor changes in land surface elevation in proglacial areas. Over coming years, the use of UAVs in glacial environments is likely to continue to develop (already, most field visits to glacial environments involve the use of at least one UAV), partly driven by technological advances (e.g., extended battery life) but also by thinking of novel ways that UAVs can help address fundamental questions in glaciology.

REFERENCES

Allaart, L., Friis, N., Ingólfsson, Ó., Håkansson, L., Noormets, R., Farnsworth, W.R., Mertes, J. and Schomacker, A. 2018. Drumlins in the Nordenskiöldbreen forefield, Svalbard. *Gff 140*(2): 170–188.

Boulton, G.S. 1976. The origin of glacially fluted surfaces-observations and theory. *Journal of Glaciology 17*(76): 287–309.

Chandler, B.M., Evans, D.J. and Roberts, D.H. 2016. Characteristics of recessional moraines at a temperate glacier in SE Iceland: Insights into patterns, rates and drivers of glacier retreat. *Quaternary Science Reviews 135*: 171–205.

Chandler, M.P.B., Lovell, H., Boston, C.M., Lukas, S., Barr, I.D., Benediktsson, Í.Ö., Benn, D.I., Clark, C.D., Darvill, C.M., Evans, D.J.A., Ewertowski, M.W., Loibl, D., Margold, M., Otto, J.-C., Roberts, D.H., Stokes, C.R., Storrar, R. and Stroeven, A.P. 2018. Glacial geomorphological mapping: A review of approaches and framework for best practice. *Earth-Science Reviews 185*: 806–846.

Ely, J.C., Graham, C., Barr, I.D., Rea, B.R., Spagnolo, M. and Evans, J. 2017. Using UAV acquired photography and structure from motion techniques for studying glacier landforms: Application to the glacial flutes at Isfallsglaciären. *Earth Surface Processes and Landforms 42*(6): 877–888.

Ewertowski, M.W., Tomczyk, A., Evans, D., Roberts, D. and Ewertowski, W. 2019. Operational framework for rapid, very-high resolution mapping of glacial geomorphology using low-cost unmanned aerial vehicles and structure-from-motion approach. *Remote Sensing 11*(1): 65.

Harvey, M.C., Rowland, J.V. and Luketina, K.M. 2016. Drone with thermal infrared camera provides high resolution georeferenced imagery of the Waikite geothermal area, New Zealand. *Journal of Volcanology and Geothermal Research 325*: 61–69.

Immerzeel, W.W., Kraaijenbrink, P.D.A., Shea, J.M., Shrestha, A.B., Pellicciotti, F., Bierkens, M.F.P. and De Jong, S.M. 2014. High-resolution monitoring of Himalayan glacier dynamics using unmanned aerial vehicles. *Remote Sensing of Environment 150*: 93–103.

Lamsters, K., Karušs, J., Krievāns, M. and Ješkins, J. 2019. Application of unmanned aerial vehicles for glacier research in the Arctic and Antarctic. In *Proceedings of the 12th International Scientific and Practical Conference 131*: 135. http://journals.rta.lv/index.php/ETR/article/view/4130

Le Heron, D.P., Vandyk, T.M., Kuang, H., Liu, Y., Chen, X., Wang, Y., Yang, Z., Scharfenberg, L., Davies, B. and Shields, G. 2019. Bird's-eye view of an Ediacaran subglacial landscape. *Geology 48*(8): 705–709.

Midgley, N.G., Tonkin, T.N., Graham, D.J. and Cook, S.J. 2018. Evolution of high-Arctic glacial landforms during deglaciation. *Geomorphology 311*: 63–75.

Noh, M.J. and Howat, I.M. 2015. Automated stereo-photogrammetric DEM generation at high latitudes: Surface extraction with TIN-based search-space minimization (SETSM) validation and demonstration over glaciated regions. *GIScience & Remote Sensing 52*(2): 198–217.

Pearce, D.M., Ely, J.C., Barr, I.D. and Boston, C.M. 2017. Glacier reconstruction. In: Cook, S.J., Clarke, L.E. and Nield, J.M. (eds.) *Geomorphological Techniques*. London: British Society for Geomorphology, 1–16.

Sieberth, T., Wackrow, R. and Chandler, J.H. 2016. Automatic detection of blurred images in UAV image sets. *ISPRS Journal of Photogrammetry and Remote Sensing 122*: 1–16.

Storrar, R., Ewertowski, M., Tomczyk, A.M., Barr, I.D., Livingstone, S.J., Ruffell, A., Stoker, B.J. and Evans, D.J. 2019. Equifinality and preservation potential of complex eskers. *Boreas*. In press.

Tang, L. and Shao, G. 2015. Drone remote sensing for forestry research and practices. *Journal of Forestry Research 26*(4): 791–797.

Tonkin, T.N. and Midgley, N.G. 2016. Ground-control networks for image based surface reconstruction: An investigation of optimum survey designs using UAV derived imagery and structure-from-motion photogrammetry. *Remote Sensing 8*(9): 786.

Tonkin, T.N., Midgley, N.G., Cook, S.J. and Graham, D.J. 2016. Ice-cored moraine degradation mapped and quantified using an unmanned aerial vehicle: A case study from a polythermal glacier in Svalbard. *Geomorphology 258*: 1–10.

Turner, D., Lucieer, A. and Wallace, L. 2013. Direct georeferencing of ultrahigh-resolution UAV imagery. *IEEE Transactions on Geoscience and Remote Sensing 52*(5): 2738–2745.

Westoby, M.J., Brasington, J., Glasser, N.F., Hambrey, M.J. and Reynolds, J.M. 2012. Structure-from-Motion photogrammetry: A low-cost, effective tool for geoscience applications. *Geomorphology 179*: 300–314.

4 Laser Scanning of a Complex Cave System during Multiple Campaigns
A Case Study of the Domica Cave, Slovakia

*Ján Kaňuk, Jozef Šupinský, John Meneely,
Zdenko Hochmuth, Ján Šašak, Michal
Gallay, and Marco Callieri*

INTRODUCTION

Caves are natural sub-surface hollow forms with an extremely complex three-dimensional (3D) morphology in both horizontal and vertical directions. Their research provides valuable knowledge for geology, hydrology, geomorphology, biology, and also history. Caves have attracted people's attention since ancient times when prehistoric humans sought refuge from adverse weather and cold or wild animals and enemies. Today, the inherent mystery and natural beauty of caves captivate human curiosity, and many have become tourist attractions that can also lead to protection. From a scientific point of view, caves are an important source of information about past environments and are important for understanding contemporary conditions and changes. Past climates can be reconstructed from the preserved natural materials, such as sediments, ice, and geomorphological forms, but also from objects of human origin (e.g., bones, working tools, paintings, ash).

From a mapping perspective, caves are a major challenge due to the complexity of surface shapes, confined spaces, lack of light, and the abundance of water, mud, or even ice (Gallay et al., 2015). Traditionally, cave surveying was mostly carried out by volunteer caving clubs and associations and, to a lesser extent, by professional cavers employing mine surveying methods, such as tacheometry. Despite the immense effort of cave surveyors, the resulting maps are highly generalized 2D floor plans or projected vertical side views, with very little 3D information. Currently, mapping with a laser distance measurer, inclinometer, and compass is widely used. Tourist 'show caves'—which have been made accessible to the general public by

DOI: 10.1201/9780429327575-4

guided visits—are usually mapped with a total station. These maps usually comprise a traverse, showing the course of the cave to which other measurements are connected—typically, the position of side corridors, passages, large speleothems, water streams, lakes, or abysses are only recorded, with little or no information on small-scale features.

Other technologies based on underground global navigation system (U-GPS) (Wenger, 2004), sonar (Stipanov et al., 2008), ground-penetrating radar (Chamberlain et al., 2000), seismic (Beres et al., 2001), or electric resistivity methods (Peterson and Berg, 2001) have previously been used to try and refine the mapping of caves.

However, these methods still are not extensively applied, principally for their complexity and demands on technical equipment and data processing. Recently, remote sensing technologies, such as close-range photogrammetry (Triantafyllou et al., 2019) and 3D laser scanning (Mohammed Oludare and Pradhan, 2016), or their combined use (Lerma et al., 2010) have become popular in cave mapping. These methods are capable of capturing an unprecedented level of detail and are faster than other methods used to date. Both these methods generate millions of 3D point measurements (usually referred to as point clouds), representing the mapped surface highly accurately in the order of millimeters and without the need to generate a surveying traverse. The application of close-range photogrammetry is, however, limited by suitable illumination usually requiring powerful artificial lights. In the case of laser scanning, which uses light detection and ranging technology (LiDAR), this darkness of caves is not a problem. For this reason, terrestrial laser scanners (TLS) have been increasingly used in mapping caves despite its relatively high cost when compared to digital cameras and lights needed for photogrammetry.

A distinction has to be made between TLS and mobile laser scanning (MLS). TLS is performed from static ground-based platforms, usually placed on a tripod, with a laser scanner rotating around its axis. It records the horizontal and vertical angle of the emitted laser beam, the time it takes that beam to return from a surface to the scanner, plus how much of that emitted beam returned, and it can do this millions of times per second. From this it can calculate a x,y,z coordinate, relative to the scanner and assign an intensity value (I) to each point calculated from the return strength of the laser from a surface.

Once a single scan is complete, it has to be moved to another location to capture the entire scene without data shadows during mapping. The individual point cloud collected from each scanning position is then registered (joined together) to generate a single point cloud in a common coordinate system.

In order for the registration to be accurate and successful, a sufficient portion of successive scans must overlap. This registration of the data is performed in dedicated, usually vendor-specific software, either manually or automatically. In recent years, a significant trend in TLS is the transfer of completed scans via a Wi-Fi or Bluetooth to a laptop or dedicated tablet for instant, in-field registration. This has many advantages in cave mapping—primarily the ability to ensure that no areas have been missed before leaving a difficult to access area.

The principle of MLS is based on the recordings of two synchronized devices: the inertial measurement unit (IMU) and the laser scanner. The IMU records the orientation angles along the x, y, z axes in 3D space to determine the trajectory of the laser scanner. The coordinates of the individual points recorded by the laser scanner are then calculated based on the laser triangulation (pulse emission angle and distance) with respect to the position of the scanner. The main benefit of MLS is in the speed of mapping and reduced data shadows by the continual movement of the scanner in space, providing an opportunity to scan around any object. On the other hand, the limited ability to record the trajectory of the scanner's motion using the IMU compromises the accuracy of the 3D coordinates of the resulting point cloud. IMU locates itself by employing various sensors, such as a gyroscope, accelerometer, compass, barometer, and sometimes GNSS. The basic problem is that the frequency of recordings by IMU is over 1,000 times lower compared to the frequency of laser scanning. These shortcomings in tracking the scanner motion must be compensated for by calculations, for example, using the simultaneous location and mapping (SLAM) method. This method is becoming widely used in MLS.

TLS is often preferred over MLS for its higher positional accuracy and high spatial density of recorded points in mapping complex cave morphologies (Mohammed Oludare and Pradhan, 2016). Although there is a wide range of less costly surveying methods, LiDAR has the potential to replace traditional techniques for cave mapping. The capabilities of TLS in cave mapping are demonstrated in this chapter by showing the results from mapping over 5,000 m underground of the World Heritage Site Domica Cave in Slovakia.

REVIEW OF THE PUBLISHED WORKS ON LASER SCANNING IN CAVES

The application of laser scanning in caves dates back to the late 1980s and only focused on renowned sites, such as Altamira in northern Spain between 1988 and 2001 (Donelan, 2002) or Cosquer Cave in France in 1994 (Thibault, 2001), which were mapped by short-range (2 m), time-consuming and laborious active triangulation scanners. Today, this scanning approach is predominantly used for high-detail 3D scanning of small objects, using commercially available devices such as Kinect (Hämmerle et al., 2014) or the FARO Freestyle™.

Even after more than 30 years since the first TLS in caves, few cave systems are mapped in a large scale. An overview of works focused on mapping caves using TLS is presented in Gallay et al. (2015) or Mohammed Oludare and Pradhan (2016). A large number of cave laser scanning projects remain unpublished or published in local magazines making them difficult to find.

Table 4.1 presents a chronological overview of publications demonstrating the use of TLS in various caves, the reason for mapping, the length of the mapped parts, and the scanning equipment used. This list is by no means a complete overview. The simple analysis of the number of scientific papers in the Scopus database by using the query (TITLE-ABS-KEY (lidar AND cave)) OR (TITLE-ABS-KEY (laser AND scanning AND cave)) found 282 documents published since 1995. This indicates the growth in the use of laser scanning to survey caves has intensified since 2008 from about 5 up to 30 publications per year in 2020.

The review in Table 4.1 suggests that, before 2010, laser scanning of caves was performed mainly for archaeological research in small but significant sites where

TABLE 4.1
Summary of Published Works Concerning Laser Scanning in Caves

Year	Author	Location	Country	Purpose of Laser Scanning Mission	Range	Type of Scanner Device
1999	Perperidoy et al. (2010)	Chapel's Cave	USA	Documentation	Unknown	Unknown
2001	Thibault	Cosquer Cave (1994)	France	Archeology	Unknown	SOISIC
2001	Robson-Brown et al.	Dordogne Caves	France	Archeology	2 scans (wall)	Surveyor ALS
2002	Donelan	Altamira Cave	Spain	Archeology	Unknown	Minolta VI-700
2003	Caprioli et al.	Castellane Grotte Cave	Italy	Archeology	100 m	Mensi-GS100
2003	Westerman et al.	Peak Cavern Vestibule	UK	Archeology	Unknown	RIEGL LMS-Z360
	El-Hakim et al.	Baiame Cave	Australia	Archeology	Unknown	RIEGL LMS-Z210i
2004	The Courier (Channel 4 Time team)	Wemyss Caves	Scotland	Archeology	Unknown	Unknown
2005	Aujoulat	Veilmouly Cave (1994)	France	Archeology	Unknown	SOISIC
2005	Fryer et al.	Baiame Cave	Australia	Archeology	Unknown	RIEGL LMS-Z210i
2005	Murphy et al.	Gaping Gill Cave	UK	Documentation	Unknown	RIEGL LMS-Z210i
2006	Beraldin et al.	Grotta dei Cervi	Italy	Archeology	Unknown	Big Scan prototype
2006	Doering et al.	Preacher's Cave	Bahamas	Archeology	20 m	Leica HDS 3000
2007	Tsakiri et al.	Kefala Cave	Greece	Documentation	Unknown	iQsun 880HE80
2008	Brich et al.	High Pesture Cave	UK	Documentation	Unknown	Trimble GS200
2008	Canavese et al.	Naica Cave	Mexico	Geology	110 m	FARO CAM2 Focus 3D
2009	Buchroithner & Geiseckner	Dachstein Southface Cave	Austria	Documentation	100 m	RIEGL LMS-Z420i
2009	Gonzalez-Aguilera et al.	Las Caldas, Pena de Candamo Caves	Spain	Archeology	Unknown	Trimble GS200
2009	Chandelier & Roche	Tautavel Cave	France	Paleontology	Unknown	Trimble GS200
2009	Pucci & Marambio	Olerdola Cave	Spain	Archeology	Unknown	RIEGL LMS-Z420

(Conitnued)

TABLE 4.1 *(Continued)*
Summary of Published Works Concerning Laser Scanning in Caves

Year	Author	Location	Country	Purpose of Laser Scanning Mission	Range	Type of Scanner Device
2009	Rüther et al.	Wonderwerk Cave	South Africa	Archeology	Unknown	Leica HDS3000
2010	Grussenmeyer et al.	Les Fraux Cave	France	Archeology	Unknown	FARO Photon 120
2010	Lerma et al.	La Cova del Parpallo Cave	Spain	Archeology	Unknown	FARO LS 880HE
2010	McIntire	Mushpot Cave	USA	Documentation	150 m	Leica HDS6000
2011	Addison	Mammoth Cave	USA		4 000 m	
2011	Buchroithner	Eisriesenwelt Cave	Austria	Cryomorphology	1 000 m	FARO Photon 120/20 (2011); FARO Focus 3D (2013)
2011	Canavese et al.	Santa Barbara Cave System	Italy	Geomorphology	740 m	Leica HDS6100 and RIEGL LMS-Z210i
2011	Jaillet et al.	Orgnac's Cave	France	Documentation	Unknown	Leica HDS 6000
2011	Petters et al.	Eisriesenwelt Cave	Austria	Cryomorphology	1 000 m	FARO Photon 120
2011	Roncat et al.	Marchenhohle Cave	Austria	Morphogenetic	150 m	Z+F Imager 5006i
2012	Azmy et al.	Gua Kelawar Cave	Malaysia	Zoology	14 scans	FARO Photon 120
2012	Buchroithner	Niah Caves	Malaysia	Documentation	Unknown	FARO Focus 3D
2012	Gašinec	Dobšinská Ice Cave	Slovakia	Cryomorphology	Unknown	Leica ScanStation C10
2012	Kordić et al.	Kuca Cave	Croatia	Archeology	Unknown	FARO Photon 120
2012	Lyons-Baral	Coronado Cave	USA	Hazards evaluation	200 m	Leica ScanStation C10
2012	Milius & Petters	Eisriesenwelt Cave	Austria	Cryomorphology	1 000 m	FARO Photon 120
2012	Santos Delgado et al.	El Sidrón Cave	Spain	Paleontology	50 m	Leica ScanStation C10
2013	Canevese and Tedeschi	Re Tiberio Cave	Italy	Documentation	60 m	Leica HDS6100

TABLE 4.1 *(Continued)*
Summary of Published Works Concerning Laser Scanning in Caves

Year	Author	Location	Country	Purpose of Laser Scanning Mission	Range	Type of Scanner Device
2013	Gede et al.	Pálvölgy Cave	Hungary	Documentation	Unknown	FARO Focus 3D, Leica ScanStation C10
2013	Lindgren & Galeazzi	Las Cuevas Cave	Belize	Documentation	Unknown	FARO Focus 3D
2013	McFarlane et al.	Gomatong Caves	Malaysia	Documentation	1 000 m	FARO Focus 3D
2013	Nash & Beardsley	Cathole Cave	Wales	Documentation	Unknown	Leica HDS6000
2013	Núñez et al.	Can Sadurní Cave	Spain	Archeology	Unknown	RIEGL LMS-Z420i
2013	Plan et al.	Mammuthöhle Cave	Austria	Geomorphology	200 m	Z+F Imager 5006i
2013	Puchol et al.	Pastora Cave	Spain	Archeology	Unknown	FARO Photon 120
2013	Silvestre et al.	Algar do Penico Cave	Portugal	Documentation	80 m	Leica ScanStation C10
2013	Yumin	Lianhua Cave, Tianlongshan Cave	China	Archeology	Unknown	Unknown
2014	Berenguer-Sempere et al.	Castil Ice Cave	Spain	Cryomorphology	72 m	Leica ScanStation C10
2014	Burens et al.	Les Fraux Cave	France	Archeology	430 m	FARO Photon 120, FARO Focus 3D
2014	Cosso et al.	Arma Pollera Cave	Italy	Documentation	Unknown	Z+F Imager 5010
2014	Hämmerle et al.	Dechen Cave	Germany	Comparsion	Unknown	RIEGL VZ-400, Kinect
2014	Hobléa et al.	Orgnac's Cave, Chauvet Cave	France	Documentation	Unknown	Leica HDS 6000
2014	Hoffmeister	Sodmein Cave	Egypt	Archeology	Unknown	RIEGL LMS-Z420i
2014	Kukutsch et al.	Amatérska Cave	Czechia	Documentation	1 300 m	Leica ScanStation C10

(Conitnued)

TABLE 4.1 *(Continued)*
Summary of Published Works Concerning Laser Scanning in Caves

Year	Author	Location	Country	Purpose of Laser Scanning Mission	Range	Type of Scanner Device
2014	Leonov et al.	Denisova Cave	Russia	Documentation	37 scans	FARO Focus 3D
2014	Novaković	Škocjan Caves	Slovenia	Documentation	Unknown	Leica ScanStation C10
2014	Tyree	Skoteino Cave	Greece	Documentation	Unknown	RIEGL LMS-Z420i
2014	Zlot & Bosse	Jenolan Caves	Australia	Documentation	17 100 m	Hannibal, Zebedee
2015	Bella et al.	Dupnica Cave	Slovakia	Geology	Unknown	Leica ScanStation C10
2015	Gallay et al.	Domica Cave	Slovakia	Geomorphology	1 600 m	FARO Focus 3D
2015	Marisco et al.	Santa Croce Cave	Italy	Documentation	90 m	Leica HDS 3000
2015	McFarlane et al.	Gomantong Caves	Malaysia	Zoology	1 000 m	FARO Focus 3D
2015	Santagata et al.	Grotta della lucerna Cave	Italy	Documentation	Unknown	Leica HDS 7000
2016	Hoffmeister	Ardelas Cave	Spain	Documentation	Unknown	RIEGL LMS-Z420i
2016	Kruger et al.	Rising Star Cave	South Africa	Archeology	Unknown	FARO Focus 3D
2016	Tyszkowski et al.	20 Caves	Poland	Documentation	Unknown	RIEGL VZ-400
2016	Yakar et al.	"Hadim" Cave	Turkey	Documentation	13 scans	OPTECH ILRIS
2017	Basantes et al.	Elviandi Cave	Ecuador	Documentation	450 m	FARO Focus 3D
2017	Citton et al.	Grotta della Básura Cave	Italy	Paleonthology	Unknown	Leica ScanStation 2
2017	Fabbri et al.	Grotta A Cave	Italy	Geology	Unknown	FARO CAM2 Focus 3D
2017	Pukanska et al.	Belianska Cave	Slovakia	Documentation	Unknown	Leica ScanStation C10
2018	De Waele et al.	Ca'Castellina Cave	Italy	Geomorphology	Unknown	FARO CAM2 Focus 3D
2018	Gómez-Lende & Sánchez-Fernández	Picos de Europa Ice Caves	Spain	Cryomorphology	Unknown	Leica ScanStation C10, FARO Focus 3D

TABLE 4.1 *(Continued)*
Summary of Published Works Concerning Laser Scanning in Caves

Year	Author	Location	Country	Purpose of Laser Scanning Mission	Range	Type of Scanner Device
2018	Petrović et al.	Pećura and Zamna Caves	Serbia	Documentation	Unknown	Leica Nova MS50
2019	Aiello et al.	Grotta dei Pipistrelli	Italy	Documentation	72 scans	FARO Focus S70
2019	Kregar et al.	Kumik Cave	Slovenia	Documentation	2 000 m	Leica BLK360
2019	Nocerino et al.	Grotta Giusti	Italy	Documentation	Unknown	Leica HDS7000
2019	Radicioni et al.	Frasassi Caves	Italy	Documentation	Unknown	FARO Focus 3D
2019	Shults et al.	Kyiv Pechersk Lavra Caves	Ukraine	Documentation	Unknown	Leica ScanStation
2019	Sorrioux et al.	Gouffre Georges	France	Geology	250 m	RIEGL VZ-1000
2019	Šupinský et al.	Silická ľadnica Cave	Slovakia	Cryomorphology	50 m	RIEGL VZ-1000
2019	Šupinský et al.	Domica Cave	Slovakia	Documentation	6 000 m	FARO Focus 3D, RIEGL VZ-1000
2019	Zeid et al.	Fumane Cave	Italy	Archeology	Unknown	Leica ScanStation C10

a small number of scan positions was sufficient (up to 10) (Robson-Brown et al., 2001; Westerman et al., 2003; González-Aguilera et al., 2009). Since 2010, laser scanning has been applied on a larger scale. The variety of scanners on the market and improved capabilities, lower price, and new methods of processing point clouds stimulated the application also in caves. Gradually, longer parts of caves were mapped, and larger areas of caves with a larger number of scanning positions were performed. The main purpose was in cave documentation (Petters et al., 2011; Kuda et al., 2014; Zlot and Bosse, 2014; Kregar et al., 2019) and geomorphological analysis (Roncat et al., 2011; Buchroithner et al., 2012; Bella et al., 2015; Fabbri et al., 2017; Gallay et al., 2016). Other applications include, for example, analysis of cryomorphological characteristics of cave ice (Gašinec et al., 2012; Gómez-Lende and Sánchez-Fernández, 2018), as well as evaluation of the volume change of glacial glaze (Milius and Petters, 2012; Šupinský et al., 2019). TLS in caves was also used in zoology for animal counting purposes, in assessing potential natural risks, and in paleontology (Azmy et al., 2012; Lyons-Baral, 2012; Citton et al., 2017).

There are two main types of TLS technology. The most widely used scanners are based on emitting pulses of laser energy (time-of-flight scanners), which reach a longer range than the second type based on continuous emission of laser energy (continuous wave (CW), phase-based scanners). In narrow passages, however, the advantage of pulse-based laser scanners cannot be fully exploited. The deployed devices comprise pulse-based scanners RIEGL LMS/VZ Series™ (Núñez et al., 2013; Tyszkowski et al., 2016; Šupinský et al., 2019), Leica ScanStation (Pukanska et al., 2017; Gómez-Lende and Sánchez-Fernández, 2018; Zeid et al., 2019), Leica BLK™ Series (Kregar et al., 2019); or phase-based scanners FARO Focus 3D X/S Series™ (Gallay et al., 2015; Aiello et al., 2019; Radicioni et al., 2019), Leica HDS™ Series (Marsico et al., 2015; Santagata et al., 2015; Nocerino et al., 2019), and Z+F IMAGER™ (Roncat et al., 2011; Plan et al., 2013; Cosso et al., 2014). The list of reviewed works supports that TLS in caves is possible even in challenging conditions (Buchroithner and Gaisecker, 2009). In addition, MLS brings new possibilities for cave mapping (Bosse et al., 2012; Zlot and Bosse, 2014; Kaul et al., 2016).

LASER SCANNING OF THE DOMICA CAVE

This case study concerns laser scanning of the Domica Cave and some of its exterior, surface surroundings. The resulting geodatabase allows for the creation of detailed and accurate maps, cross-sections, and plan views and also extends geomorphological, climatological, speleo-biological, and hydrological research of the cave. TLS has been carried out on the cave system from 2014. First, the show cave part was scanned (ca. 1,500 m), and then other publicly unavailable parts followed. In 2014, an airborne laser scanning (ALS) campaign was carried out to map the surface.

Geographical Setting

Domica Cave is located in the Triassic limestones of the Slovak Karst Mountains in southeastern Slovakia, about 1 km west of the border with Hungary (Figure 4.1). The cave is part of a much longer system continuing through the state border into the Aggtelek Karst, where it is called Baradla Cave. The Domica-Baradla Cave system has a total length of 27,476 m (Gaál and Gruber, 2014). The Slovakian (Domica) section has a length of 8,014 m. It is a unique cave with colorful flowstone decoration characterized by cascading lakes, typical onion-shaped stalactites, flowstone drums, shields, and stegamites. The uniqueness of the cave is also emphasized by specific fauna and flora. For these reasons, the cave is a listed UNESCO World Natural Heritage Site and a Ramsar Site. The detailed characteristics of the entire Domica-Baradla Cave system can be found in Gaál and Gruber (2014).

The cave was formed by underground streams, two of which still flow through the system: the Domický stream and the Styx river, which continues into the Hungarian section. Domica Cave is regularly affected by floods in winter. These floods have a destructive effect on the decoration of the cave as well as on the infrastructure, especially in the publicly accessible areas. The presence of tourists has also induced anthropogenic interventions in the cave such as building water dams and pavements. A better understanding of the reoccurring floods were one of the rationales behind the application of TLS and ALS to generate a detailed 'digital twin' of this system.

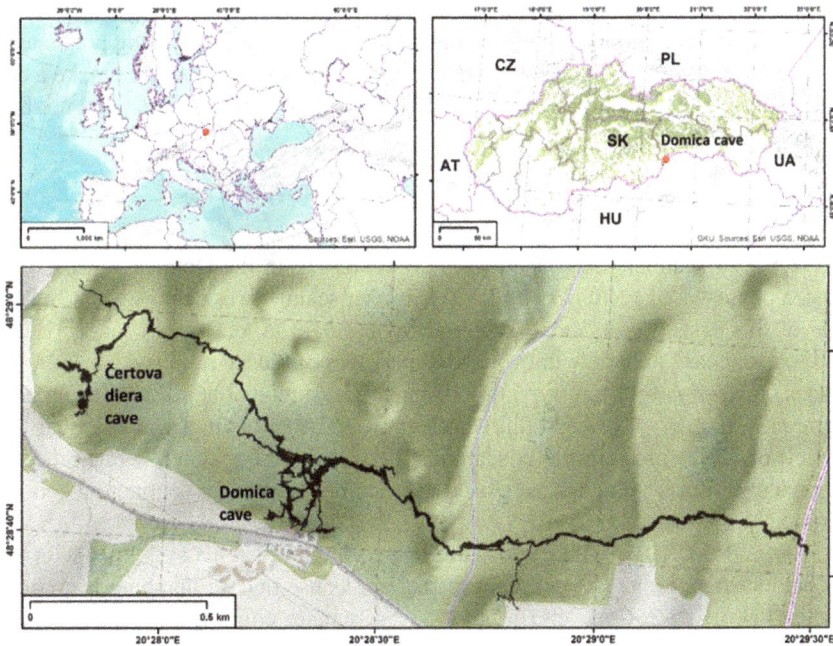

FIGURE 4.1   Location of the Domica cave system including the Devil's Hole Cave (Čertova diera) overlayed with shaded airborne LiDAR digital terrain model and land cover map.

Source: © Open Street Map

Domica Cave has been mapped several times since its discovery in 1932 by a soldier, Ján Majko. Traditionally mine-surveying methods and equipment were used for mapping. Just after the discovery, the first comprehensive mapping of Domica was supervised by mining surveyor Paloncy. The aim of this survey was to generate a map of the new cave in the territory of former Czechoslovakia. At that time, the cave contained prehistoric artifacts untouched since the Neolithic people left the cave—thought to be due to the collapse of the entrance over 5,000 years ago. In 1937, Roth carried out a much more detailed mapping exercise, which focused on large halls rich in cave decoration. This produced a detailed map of selected parts of the cave at a scale of 1:100.

The purpose of surveying the cave in 1949 was in locating and establishing the underground state border between Slovakia and Hungary. The mission resulted in a highly accurate and well-stabilized surveying network with detailed recording of the measurements. In 1964, a map created by Droppa and Chovan was published (Droppa 1964), complementing the 1949 cave floor plan with side elevations. Further mapping by Droppa (1972) recorded the ground levels of the cave passages, and this indicated a gradual erosion of the cave base during its formation. The successive opening of new parts of the cave led to greater invasive interventions. In order to build an artificial tunnel from Suchá chodba to the Panenská chodba passage, a detailed mine survey was carried out in 1975 (Novoveský, 1975). The surveying points used in all these surveys are still present and can be used to connect new surveys with a high degree of accuracy. Hochmuth (2014) linked his survey to these existing points using traditional mine-surveying techniques. The aim of this campaign was to make a continuous traverse through the cave and extend the survey into areas that had not been measured before.

LASER SCANNERS USED IN MAPPING THE CAVE

Two types of scanners were used for TLS surveying in the Domica Cave system. First, a phase-based FARO Focus 3D X 130 laser scanner was deployed. The advantage of this system is its small size and light weight (5 kg), which makes it very portable and easy to handle in the narrow passages of the cave. This device scans at ranges between 0.6–130 m providing distance measurement at ± 2 mm using near-infrared laser energy of 1,550 nm wavelength. Its benefit is in a wide vertical field of view of 310°, which allows for scanning areas above the scanner. White reflective spheres of uniform diameter were used as reference targets for semiautomated registration of the scans.

From 2015, a RIEGL VZ-1000 scanner was employed. This system is primarily for outdoor long-range surveying, nevertheless, its use in cave mapping is common (Table 4.1). It is a full waveform pulse-based scanner emitting near-infrared laser pulses of 1,550 nm wavelength. The measurement precision along the range direction is ± 3 mm with a minimal scanning range from 1.5 m to 1,400 m. Compared to the FARO scanner, the VZ-1000 is relatively heavy (10 kg with batteries), which made it difficult to handle in the cave. The most significant drawback of this scanner in caves is the limited vertical angle of 100°, which complicates capturing data on the ceiling directly above the scanner. The data shadows created with this system have to be reduced by closer placement of scanning positions to each other or scanning the ceiling by tilting the scanner.

Besides the distance between consecutive scanning positions, the density of point measurements is controlled by the frequency of measurements. The RIEGL VZ-1000 is capable of emitting 550,000 pulses per second (550 kHz pulse repetition rate). Scanning at this rate takes approximately 80 seconds with a scanning detail of 0.06° in the vertical and horizontal directions. The measurement frequency of the FARO Focus 3D X 130 is up to 950,000 points per second and at resolution ¼ (0.036°), scanning from one position takes approximately 3 minutes and 26 seconds, although this time can be increased to improve the quality of the data. The VZ-1000 is capable of recording an unlimited number of pulse echoes. Practically, only echoes above a set quality threshold are recorded, and they can be further filtered based on pulse waveform deviation or rescaled intensity. This information can be used to remove stray points.

THE WORKFLOW OF LiDAR CAVE MAPPING

The TLS data acquisition is followed by several steps of data processing. The main steps of the workflow to generate a cave map are shown in Figure 4.2. The first task is the registration of individual scans acquired from each scanning location. The

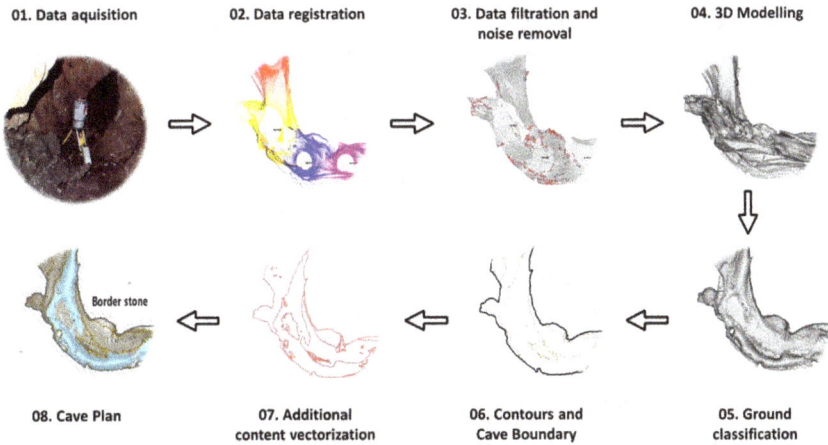

FIGURE 4.2 The workflow of converting the TLS point cloud into a cave map.

first two steps in Figure 4.2, data acquisition and registration, are the most time-consuming. Scanning positions need to be carefully chosen to keep their number reasonably low but, at the same time, minimize data shadows while still ensuring a sufficient overlap with the previous scan. Ultimately, a carefully planned TLS survey allows for the application of automated registration procedures for adjusting the scan positions, resulting in minimized registration errors, typically in the order of millimetres. Gallay et al. (2015) explain the details of the TLS methodology of scanning in 2014 and Šupinský et al. (2019) describe the following TLS campaigns.

After the scans are registered into a single point cloud, it is necessary to filter out erroneous data (Kaňuk et al., 2019). These are mainly stray points form laser reflections with water or points with a high position of uncertainty due to low surface reflectivity. Point filtering and the denoising process are prerequisites for deriving any complex and realistic 3D surface models.

THE MULTIFOLD CAVE LASER SCANNING CAMPAIGN

Data collection with a TLS was carried out in 43 separate surveys, totaling 178 scanning hours. From a practical point of view, the most time-consuming factor was ensuring the safe transport of the surveying equipment to a location and the stabilization of the device in very narrow passages in areas with water and mud. To date, this mapping has involved 1,029 scan locations with the RIEGL VZ-1000 and 786 positions with the FARO Focus 3D, generating an average of 9 million points per position. A typical day involved 4–8 hours of scanning underground, 46 scan locations with a spacing ranging from 2 to 20 m to ensure sufficient scan overlap. This achieved a point density ranging from 26,000 points.m^{-2} in large domes to 46,000 points.m^{-2} in narrow passages.

The first phase of mapping was performed in March 2014 with the FARO Focus 3D scanner (Gallay et al., 2015). This involved 327 positions during a 5-day campaign which primarily focused on the show cave part with relatively easy access (Figure 4.3 (C)) and several narrow passages with the base covered by dry clay or limestone. Approximately 1,600 m of underground corridors were mapped including a section of the aboveground visitor's entrance to the cave system. The period from 2014 to 2017 focused on data processing and 3D modeling of the cave surface from this survey (Gallay et al., 2016; Hofierka et al., 2017). A systematic extension of this initial survey using the RIEGL TLS into other parts of the cave started in 2017. Initially, it focused on scanning the parts with large domes and corridors, and due to the low water level of the underground Styx river, the riverbed could have been mapped along its entire length. It was linked to the survey from 2014 by scanning a large

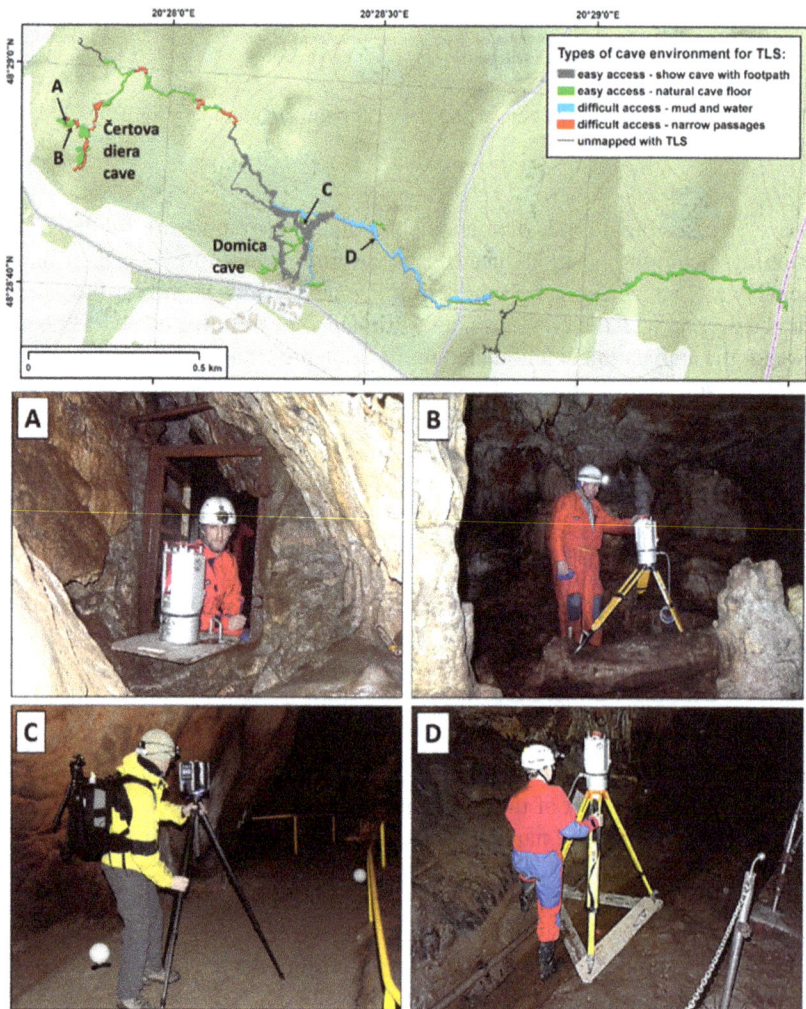

FIGURE 4.3 Various kinds of environmental conditions while laser scanning the Domica Cave are annotated with letters (A–D) and located on the map.

overlapping area that contained an artificial dam. There was no water present at the time of this survey, but the bottom of the riverbed was covered by mud. To prevent the scanner from sinking into this mud, a platform was made from wooden boards, which stabilized it during data collection (Figure 4.3 (D)). This allowed sections of the cave right up to the border with Hungary to be surveyed.

In 2018, the TLS mapping focused mainly on the Čertova diera Cave (Devil's Hole). This part of the system is characterized by alternating spacious domes and narrow passages in which one has to crawl to pass through. The dimensions of these narrow corridors would not allow the scanner to be placed on a tripod; therefore, to scan these parts, a steel platform was constructed on which the scanner was mounted (Figure 4.3 (A)). TLS in this part of the cave could only be carried out during periods of very low water level in the Styx river. A total of approximately 6,000 m of these very narrow corridors were surveyed by TLS. The resulting data set from this second TLS campaign contains over 25 billion points. The next phase was cleaning and filtering the data, and this represented approximately 10% of the initial point cloud (Hofierka et al., 2017).

The key procedure in TLS data post-processing is registration where at least four common points between overlapping scans need to be co-located. This task was performed in vendor-specific software—FARO Scene™ and RIEGL RiScanPro.™

White registration spheres (Figure 4.3 (C)) were used to achieve this in the first Domica TLS survey of 2014 (Gallay et al., 2015). The scans acquired in the following campaigns by RIEGL VZ-1000 (Figure 4.3 (B)) were co-registered without any artificial targets by, first, manually finding four identical points in the overlapping scans followed by an automatic orientation. Then, an automatic multi-station adjustment (MSA) procedure was used with a robust fitting mode to closely match the scans based on their area of overlap. This step resulted in finding groups of points (i.e. polydata), which represent identical parts of the scanned surface within the specified radii. By this means, the number of points used in the subsequent MSA procedure (registration) increased, providing a more accurate registration (Ullrich et al., 2003). For the coarse registration, the standard deviation ranged from 8 to 15 mm, but after subsequent iterations of MSA, the resulting standard deviation of the internal registration of positions improved to 3 mm.

The integration of the first point cloud from 2014 with the subsequent point clouds was solved by importing all individual positions scanned in 2014 into the registration project in RiSCAN Pro and registering them with the rest of the scanned data. After the first stage of registration, the MSA procedure was used to closely match the scans which formed a closed traverse loop. The first position remained fixed with all other scan locations subsequently aligned to this initial scan. When closing the scan survey loop of the 2014 data, the accumulated standard error was markedly reduced from 150 mm to 4 mm overall. The survey traverse could not be closed in the long passage from the second water dam (east of Majkov Dome) to the Hungarian border.

To generate a continuing traverse from the underground state border with Hungary to the Majkov Dome, two water dams in the show cave part had to be crossed. This was possible due to the low water level at the time of scanning. Despite these favourable conditions, the sediment was unstable for placing the scanner securely on a tripod; therefore, a wooden platform was used in these areas (Figure 4.3 (D)). This platform was also used to move the scanner by boat between some locations saving

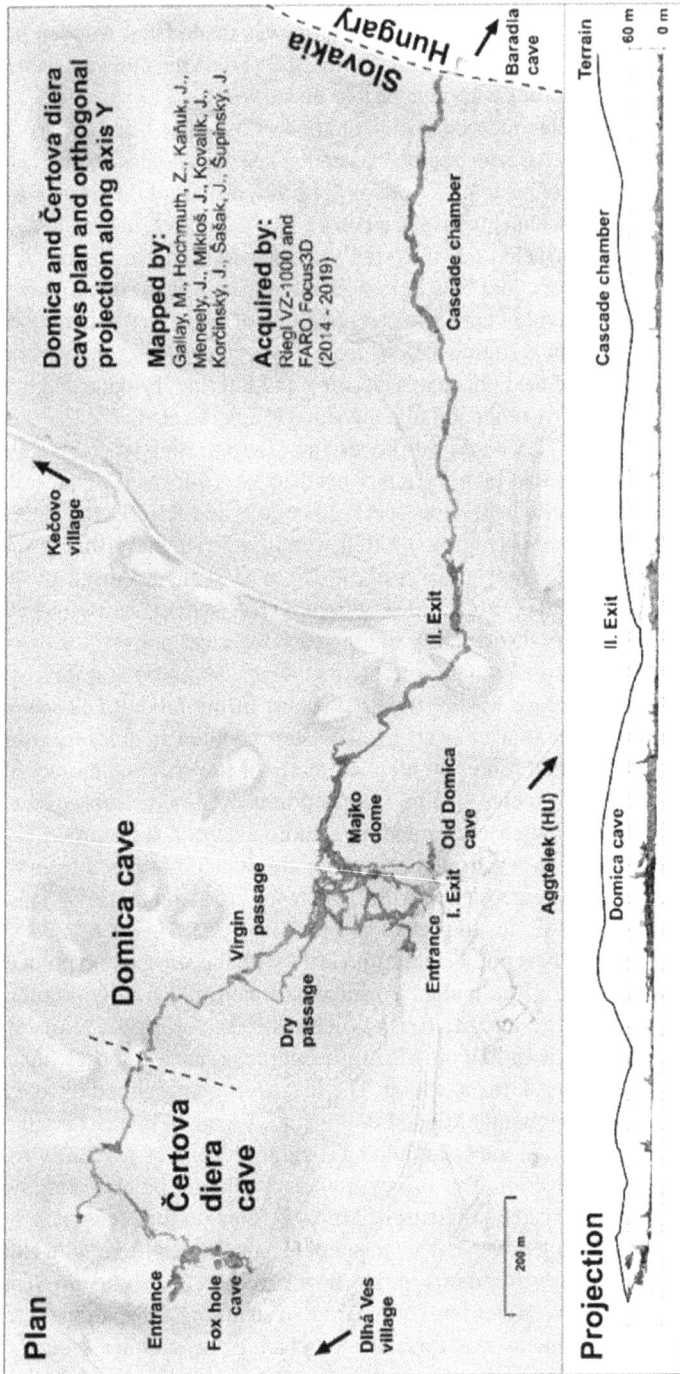

FIGURE 4.4 Top view and side view of the Domica Cave system resulting from multiple TLS campaigns combined with an airborne LiDAR digital terrain model.

valuable time in dismantling and reassembly. After scanning via the second water dam (2. plavba), the survey continued aboveground through a man-made exit from the system where control points, surveyed with an RTK-GNSS, were measured with the scanner. The registration error is at its highest in the part from the second entrance to the Hungarian border as the scanning survey traverse remains open, and the cumulated standard deviation of error was up to 200 mm. The use of artificial targets was not feasible in such extreme environmental conditions in this part of the cave; therefore, a manual selection of common points and the iterative MSA procedure was preferred as the most appropriate solution.

Redundant data, such as points from scattered reflections on sharp edges on railings, speleothems, and wet surfaces, were removed from the data set. Some mirrored features were observed when speleothems were covered with a thin film of water. Also, when scanning in winter near the cave exit, cold air mixed with the warmer air in the cave, producing a lot of noisy data on the floor and the cave ceiling. Therefore, in such environmental conditions, it is recommended to scan only when the air temperature gradient is low, ideally in light winds. After the point cloud was cleaned, its point density was unified by decimating the point cloud to a 10 mm resolution resulting in 2,000 million points. The extent of the final point cloud is shown in Figure 4.4.

Once registration was complete, the resulting point cloud was georeferenced to the national coordinate system (EPSG code: 5514 S-JTSK Křovák East North). This procedure made use of 56 points identified in the scans that were mapped by Novoveský (1975). The resulting standard deviation of transforming the data to the national coordinate system is 0.016 m. The belowground TLS cave data was supplemented with an airborne LiDAR (ALS) point cloud supplied by the company Photomap. More details on this ALS survey can be found in Hofierka et al. (2017, 2018).

Generating the 3D Cave Surface Model

The point cloud of the cave allows for precise measurements and the generation of cross-sections, plans, and visualizations. However, volume calculations and advanced analyses of the cave surface including geomorphometry or water flow modelling require the creation of a 3D digital surface model.

The key prerequisite for this task is in the calculation of normal vectors for each point to define the interior of the cave. Various approaches exist to achieve this. but for the complex cave surface morphology, the orientation of the normal is usually incorrectly determined if the normal point vectors are based on simple neighbourhood analysis. This is especially true on speleothems and various isolated geomorphological features. For this reason, normal vectors need to be oriented with respect to their scan location. After this step, the normals are correctly defined and a correct surface model of the entire cave is derived.

The 3D digital surface models (Figure 4.5) were created in the open-source software CloudCompare (Girardeau-Montaut, 2018) using the Screened Poisson surface reconstruction interpolation method (Kazhdan and Hoppe, 2013). The quality of the resulting surface depends on the presence of data voids, noise, the level of detail of the collected data, and the spatial resolution of the output model. After creating the surface model of the cave, it is necessary to extract areas of interest required for simulations.

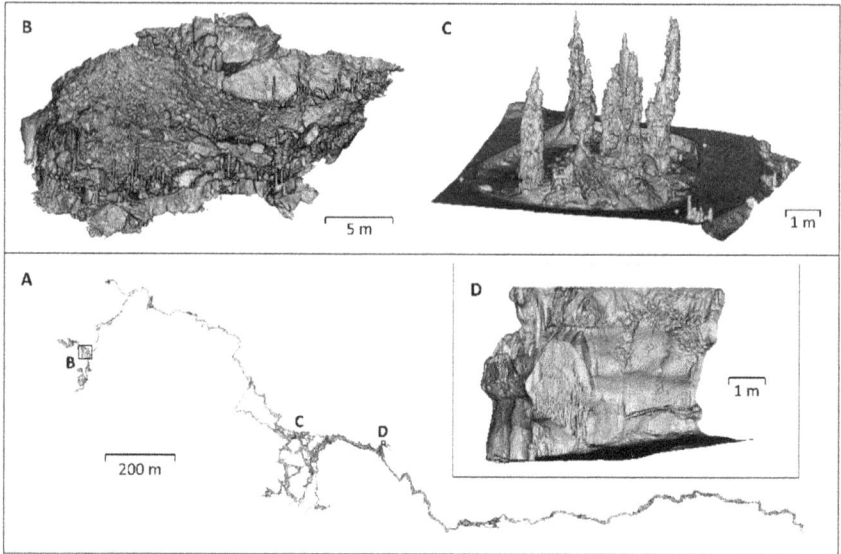

FIGURE 4.5 (A) selected parts of the 3D cave surface model, (B) showing the level of detail preserved in the model of thin ceiling stalactites, (C) massive stalagmites in the Dome of the Indian Pagodas and (D) a stegamite.

NEW MEANS OF CAVE VISUALISATION AND APPLICATION

The high level of detail and spatial extent of the TLS surveyed cave system opens new possibilities for visualising and communicating its complex geometry to researchers or the general public. Traditional speleocartography can be enhanced by including planar views from the 3D model and shading/colouring the surface with a range of attributes (e.g., rock material or a morphometric parameter) (see Figure 4.6). However, a map is still a static 2D visualisation of a 3D space, and recent developments in web-based 3D technologies have enabled interactive visualisation and analysis of large 3D point clouds. It is now possible to integrate the 3D content on the Internet directly into the browser without plug-ins or additional components. For example, Silvestre et al. (2015) presented an approach in which X3-D, WebGL, and X3-DOM were used to enable online 3D visualization and navigation of the interior of the Algar do Penico Cave, Portugal, in several different Web browsers. Potenziani et al. (2015) introduced their 3-D Heritage On-line Presenter (3-DHOP), which is an open-source software package for the creation of interactive Web presentations of high-resolution 3D models. This, in turn, enhances communication of the scientific results to a wider audience, providing improved presentation, dissemination, and further analysis (Scopigno et al., 2017).

For these reasons, a stand-alone LiDAR Web portal of the cave survey was produced. This portal was generated using the Laspublish software utility in the LAStools package (Rapidlasso, 2019). This uses the Potree open-source WebGL-based renderer (Scheiblauer, 2014; Schütz, 2016; Potree, 2021). Potree is capable of efficiently

Stará Domica

Plan view and orthogonal projections of the cave
Terrestrial laser scanning with Riegl VZ-1000
Registration StDev: 0.0036 m
Horizontal datum: S-JTSK - Krovak East North
Vertical datum: Kronstadt level Baltic height
Mapped by: Hochmuth Z., Mikloš J., Šupinský J.
Date of mapping: 27.08.2019

Legend:

ʎ	stalagmite	┬ ┬	chimney
Υ	stalactite		contour
Ι	pillar		flowstone
▲	survey		massif
∿	cave boundary		clay
──	profile		debris
┷┷┷	open pit		max. cave extent

Vertical view

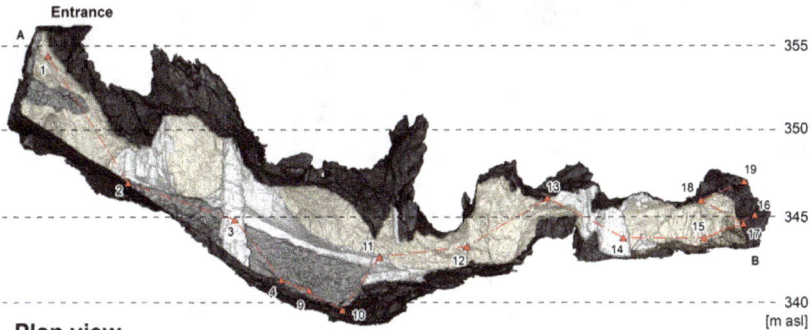

Plan view

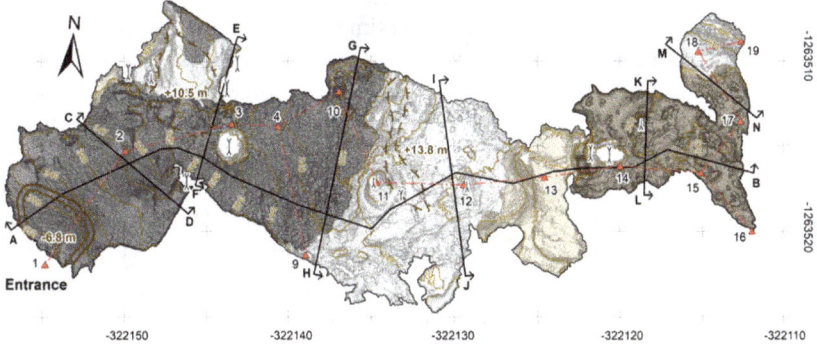

Selected profiles

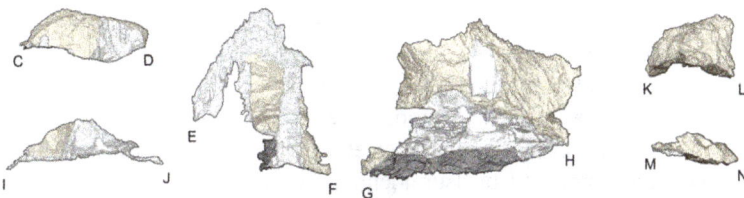

FIGURE 4.6 Example of a cave map of a part of the Domica system called Stará Domica (Old Domica, Figure 4.4) resulting from the TLS campaigns. The map contains shaded relief of the cave base with the DEM coloured according to its material.

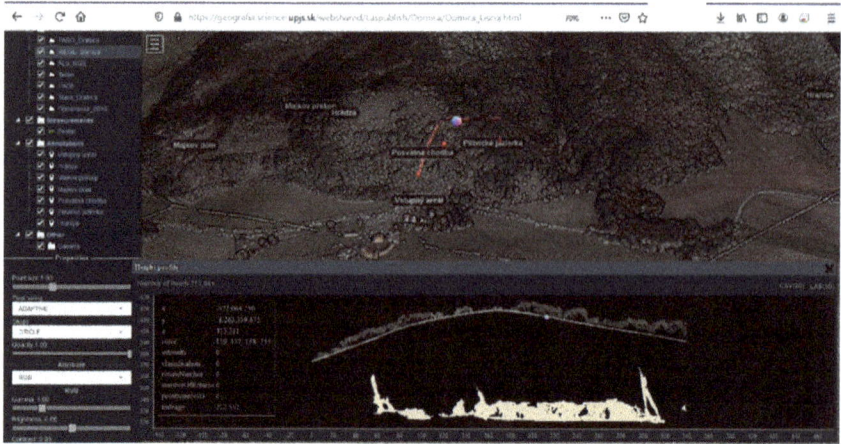

FIGURE 4.7   Interactive online 3D visualisation and analysis of the cave 3D point cloud combined with the airborne LiDAR using the Potree interface (Schütz et al., 2020).

Source: https://uge-share.science.upjs.sk/webshared/Laspublish/Domica/Domica_Liscia.html

visualising nearly 600 billion points in real time via the Internet, allowing the user to change the colour of points; perform distance, area, and volume measurements; generate vertical profiles; and export the data in various formats. Figure 4.7 demonstrates these capabilities and the previous data set can be accessed via the link: https://uge-share.science.upjs.sk/webshared/Laspublish/Domica/Domica_Liscia.html

To display the 3D cave surface and its parameters online, an interactive visualisation tool was generated using the platform of the 3-DHOP.[1] This tool and a 3D mesh of the Domica Cave are available at http://vcg.isti.cnr.it/varie/cave/ (Figure 4.8). The interface supports zooming, rotating, panning, changing the source light direction, and measuring Euclidean 3D distances between two points. The model for this site was generated in Meshlab using a reduced number of scan points (3.13 million) and with the octree depth of 13 (Gallay et al., 2016). Further use in 3-DHOP required conversion of the model into the compressed NEXUS format,[2] which is based on a multi-resolution data structure (Cignoni et al., 2005). The size of the model was reduced from 148 MB in .ply format to a 20 MB .ply version after conversion. This format allows faster streaming and smoother rendering in the browser.

The use of the TLS cave data is not solely restricted to the production of more accurate maps and improved measurements or visualisations. But it is extremely useful in other areas of speleology. Gallay et al. (2016) presents a study of ceiling channels in the Domica system by calculating 3D morphometric parameters derived from their 3D mesh. These speleoforms are extremely inaccessible at a height of 3 m to 10 m above the base of the cave. 3D modelling of these features revealed channels and provided evidence for anastomosis (a connection between tubular structures) of the Styx river as a significant process in the formation of the cave system.

Figure 4.9 is an example of simulating a real flood event in the cave system using the Delft3D FM (D-FlowFM)[3] modelling tool in which a highly detailed 3D model of the cave floor is the main input.

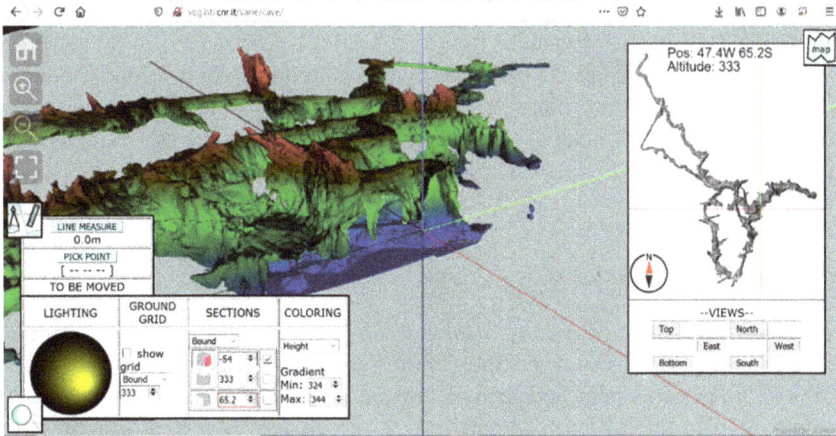

FIGURE 4.8 Interactive online 3D visualisation and analysis of the cave surface model via the 3-DHOP interface (Potenziani et al., 2015).

*Source:*http://vcg.isti.cnr.it/varie/cave/

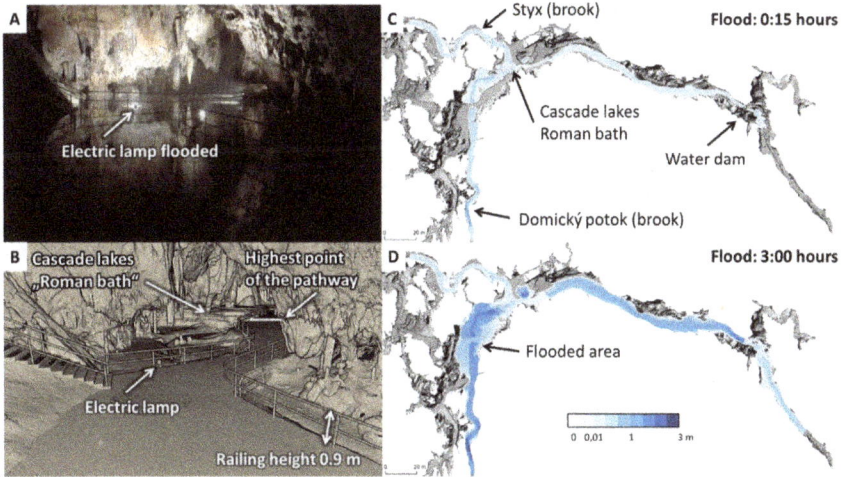

FIGURE 4.9 Cave flood modelling in the Majkov Dome (10 February 2016) with (A) a marked light source and (B) the point cloud of the dome. (C) Water depth during the simulated flood event after 15 minutes and (D) 3 hours of water inflow from the southwest and northwest (0.1 m³.s⁻¹), outflow at water dam set to 0.14 m³.s⁻¹.

CONCLUSIONS

The TLS survey of the Domica Cave system commenced in 2014 with a research project focused on developing new methods of 3D spatial modelling and surface analysis. The aim was to generate a 3D point cloud of the cave with ultra-high spatial resolution so that its surface morphology could be studied and possibly linked to aboveground geomorphology. This project resulted in the discovery of new speleo-features, proving cave anastomosis using 3D geomorphometry, and improved methods of 3D visualisation and modelling of recurring floods in the cave system. The combined TLS campaigns resulted in high-resolution mapping of over 5,000 m of underground passages. This data contributed to more accurate, informative cave maps, scientific research of inaccessible features, improved cave management, and produced new interactive visualisations accessible to all with Internet access.

ACKNOWLEDGEMENTS

The presented research originated thanks to the financial support of the Ministry of Education, Science, Research and Sport of the Slovak Republic under grant nr. VEGA 1/0168/22 'Paleogeographic and geodynamic interpretations of detrital minerals from selected areas of the Western Carpathians: a case study of the identification of the nature of transport conditions and source areas in karst and non-karst areas'.

NOTES

1. http://3-Dhop.net/
2. www.vcg.isti.cnr.it/nexus/
3. https://oss.deltares.nl/web/delft3d/home

REFERENCES

Addison, A. (2011, April). *LIDAR at Mammoth Cave*. Civil Engineering Surveyor, 22–25.
Aiello, D., Basso, A., Spena, M.T., D'Agostino, G., Montedoro, U., Galizia, M., . . . & Santagati, C. (2019). The virtual batcave: A project for the safeguard of a UNESCO WHL fragile ecosystem. *International Archives of the Photogrammetry*, Remote Sensing & Spatial Information Sciences.
Azmy, S.N., Sah, S.A., Shafie, N.J., Ariffin, A., Majid, Z., Ismail, N.A., & Shamsir, S. (2012). Counting in the dark: Non-intrusive laser scanning for population counting and identifying roosting bats. *Scientific Reports*, 2, 524. http://doi.org/10.1038/srep00524
Basantes, J., Godoy, L., Carvajal, T., Castro, R., Toulkeridis, T., Fuertes, W., . . . & Addison, A. (2017). Capture and processing of geospatial data with laser scanner system for 3D modeling and virtual reality of Amazonian Caves. In: *2017 IEEE Second Ecuador Technical Chapters Meeting (ETCM)*. http://doi.org/10.1109/etcm.2017.8247455
Bella, P., Littva, J., Pukanská, K., Gašinec, J., & Bartoš, K. (2015). Use of terrestrial laser scanning for the investigation of structural geological discontinuities and morphology of caves: On the example of the Dúpnica Cave, Západné Tatry Mts., Slovakia. *Acta Geologica Slovaca*, 7(2), 92–102.
Beraldin, J.A., Blais, F., Cournoyer, L., Picard, M., Gamache, D., Valzano, V., . . . & Gorgoglione, M. (2006). Multi-resolution digital 3D imaging system applied to the recording of grotto sites: the case of the Grotta dei Cervi. In: *VAST06: The 7th International Symposium on Virtual Reality, Archaeology and Intelligent Cultural Heritage*. The Eurographics Association.
Berenguer Sempere, F., Gómez-Lende, M., Serrano, E., & de Sanjosé-Blasco, J.J. (2014). Orthothermographies and 3D models as potential tools in ice cave studies: The Peña Castil Ice Cave (Picos de Europa, Northern Spain). *International Journal of Speleology*, 43(1), 35–43. http://doi.org/10.5038/1827-806X.43.1.4

Beres, M., Luetscher, M., & Olivier, R. (2001). Integration of ground-penetrating radar and microgravimetric methods to map shallow caves. *Journal of Applied Geophysics*, 46(4), 249–262.

Bosse, M., Zlot, R., & Flick, P. (2012). Zebedee: Design of a spring-mounted 3-d range sensor with application to mobile mapping. *IEEE Transactions on Robotics*, 28(5), 1104–1119.

Buchroithner, M.F., & Gaisecker, T. (2009). Terrestrial laser scanning for the visualization of a complex dome in an extreme alpine cave system. *Photogrammetrie-Fernerkundung-Geoinformation (PFG)*, 4, 329–339.

Buchroithner, M.F., Petters, C., & Pradhan, B. (2012). Three-dimensional visualization of the world-class prehistoric site of the Niah Great Cave, Borneo, Malaysia. In: Kremers, H. (Ed.), *Conference Handout at the Digital Cultural Heritage Interdisciplinary Conference*. Saint-Dié-des-Vosges, France, 2 p.

Burens, A., Grussenmeyer, P., Carozza, L., Leveque, F., Guillemin, S., & Mathe, V. (2014). Benefits of an accurate 3D documentation in understanding the status of the Bronze Age heritage cave "Les Fraux" (France). *International Journal of Heritage in the Digital Era*, 3(1), 179–196. http://doi.org/10.1260/2047-4970.3.1.179

Canevese, E.P., Forti, P., Naseddu, A., Ottelli, L., & Tedeschi, R. (2011). Laser scanning technology for the hypogean survey: The case of Santa Barbara karst system (Sardinia, Italy). *Acta Carsologica*, 40(1), 65–77.

Canevese, E.P., Tedeschi, R., Forti, P., & Mora, P. (2008). The use of laser scanning techniques in extreme contexts: The case of Naica Caves (Chihuahua, Mexico). *Geologia Tecnica & Ambientale (Journal of Technical & Environmental Geology)*, 2, 19–37.

Caprioli, M., Minchilli, M., Scognamiglio, A., & Strisciuglio, G. (2003). Using photogrammetry and laser scanning in surveying monumental heritage: le Grotte di Castellana. *International Archives of Photogrammetry Remote Sensing and Spatial Information Sciences*, 34(5/W12), 107–110.

Chamberlain, A.T., Sellers, W., Proctor, C., & Coard, R. (2000). Cave detection in limestone using ground penetrating radar. *Journal of Archaeological Science*, 27(10), 957–964.

Cignoni, P., Ganovelli, F., Gobbetti, E., Marton, F., Ponchio, F., & Scopigno, R. (2005). Batched multi triangulation. In: *IEEE Visualization, 2005, VIS 05*, 23–28 October 2005, Minneapolis, MN, 207–214. http://doi.org/10.1109/VISUAL.2005.1532797

Citton, P., Romano, M., Salvador, I., & Avanzini, M. (2017). Reviewing the upper Pleistocene human footprints from the 'Sala dei Misteri'in the Grotta della Bàsura (Toirano, northern Italy) cave: An integrated morphometric and morpho-classificatory approach. *Quaternary Science Reviews*, 169, 50–64.

Cosso, T., Ferrando, I., & Orlando, A. (2014). Surveying and mapping a cave using 3D laser scanner: The open challenge with free and open source software. *The International Archives of Photogrammetry, Remote Sensing and Spatial Information Sciences*, 40(5), 181.

Doering, T., Collins, L., & Branas, C. (2006). *Preacher's Cave High Definition Survey and 3D Laser Scanning Project*, Eleuthera, Bahamas: Scanning.

Donelan, J. (2002). Making prehistory. *Computer Graphics World*, March, pp. 32–33. Penn Well Publishing Co.

Droppa, A. (1964). Domica, plán 1:1000. *Mapový archív SMOPaJ*, ev. č. 16212.

Droppa, A. (1972). Príspevok k vývoju jaskyne Domica. *Československý kras*, 22, 65–72.

El-Hakim, S.F., Fryer, J., & Picard, M. (2004). Modelling and visualization of aboriginal rock art in the Baiame cave. *International Archives of Photogrammetry and Remote Sensing*, 35(5), 990–995.

Fabbri, S., Sauro, F., Santagata, T., Rossi, G., & De Waele, J. (2017). High-resolution 3-D mapping using terrestrial laser scanning as a tool for geomorphological and speleogenetical studies in caves: An example from the Lessini mountains (North Italy). *Geomorphology*, 280, 16–29.

Fryer, J.G., Chandler, J.H., & El-Hakim, S.F. (2005). Recording and modelling an aboriginal cave painting: With or without laser scanning? *International Archives of Photogrammetry, Remote Sensing and Spatial Information Sciences*, 36(5/W17), 1–8.

Gaál, Ľ., & Gruber, P. (2014). *Jaskynný systém Domica-Baradla*. Jaskyňa, ktorá nás spája. Aggtelek (Aggteleki Nemzeti park), 512.

Gallay, M., Hochmuth, Z., Kaňuk, J., & Hofierka, J. (2016). Geomorphometric analysis of cave ceiling channels mapped with 3D terrestrial laser scanning. *Hydrology and Earth System Sciences*, 20, 1827–1849.

Gallay, M., Kaňuk, J., Hochmuth, Z., Meneely, J., Hofierka, J., & Sedlák, V. (2015). Large-scale and high-resolution 3-D cave mapping by terrestrial laser scanning: A case study of the Domica Cave, Slovakia. *International Journal of Speleology*, 44(3), 277–291.

Gašinec, J., Gašincová, S., Černota, P., & Staňková, H. (2012). Zastosowanie naziemnego skaningu laserowego do monitorowania lodu gruntowego w Dobszyńskiej Jaskini Lodowej. *Inżynieria Mineralna*, 13, 31–42.

Gede, M., Petters, C., Nagy, G., Nagy, A., Mészáros, J., Kovács, B., & Egri, C. (2013). Laser scanning survey in the Pálvölgy Cave, Budapest. In: *Proceedings of the 26th International Cartographic Conference*. International Cartographic Association, Dresden, 905.

Girardeau-Montaut, D. (2018). *CloudCompare 2.10.2 Zephyrus*. www.cloudcompare.org/.

Gómez-Lende, M., & Sánchez-Fernández, M. (2018). Cryomorphological topo-graphies in the study of ice caves. *Geosciences*, 8(8), 274.

González-Aguilera, D., Muñoz, A.L., Lahoz, J.G., Herrero, J.S., Corchón, M.S., & García, E. (2009). Recording and modeling Paleolithic caves through laser scanning. In: *2009 International Conference on Advanced Geographic Information Systems & Web Services*, 19–26. IEEE.

González-Aguilera, D., Muñoz-Nieto, A., Gómez-Lahoz, J., Herrero-Pascual, J., & Gutierrez-Alonso, G. (2009). 3D digital surveying and modelling of cave geometry: Application to paleolithic rock art. *Sensors*, 9(2), 1108–1127.

Grussenmeyer, P., Landes, T., Alby, E., & Carozza, L. (2010). High resolution 3D recording and modelling of the Bronze Age cave "Les Fraux" in Périgord (France). *The International Archives of the Photogrammetry, Remote Sensing and Spatial Information Sciences*, 38, 262–267.

Hämmerle, M., Höfle, B., Fuchs, J., Schröder-Ritzrau, A., Vollweiler, N., & Frank, N. (2014). Comparison of kinect and terrestrial lidar capturing natural karst cave 3-d objects. *IEEE Geoscience and Remote Sensing Letters*, 11(11), 1896–1900.

Hochmuth, Z. (2014). Mapovanie prepojenia Čertovej diery a Domice. *Spravodaj Slovenskej speleologickej spoločnosti*, 45(3), 18–23.

Hofierka, J., Gallay, M., Bandura, P., & Šašak, J. (2018). Identification of karst sinkholes in a forested karst landscape using airborne laser scanning data and water flow analysis. *Geomorphology*, 308, 265–277.

Hofierka, J., Šašak, J., Šupinský, J., Gallay, M., Kaňuk, J., & Sedlák, V. (2017). 3D mapovanie krajiny pomocou pozemného a leteckého laserového skenovania. *Životné prostredie*, 51, 21–27.

Kaňuk, J., Zubal, S., Šupinský, J., Šašak, J., Bombara, M., Sedlák, V., Gallay, M., Hofierka, J., & Onačillová, K. (2019). Testing of V3.sun module prototype for solar radiation modelling on 3D objects with complex geometric structure. *International Archives of the Photogrammetry, Remote Sensing and Spatial Information Sciences—ISPRS Archives*, 42(4/W15), 35–40.

Kaul, L., Zlot, R., & Bosse, M. (2016). Continuous-time three-dimensional mapping for micro aerial vehicles with a passively actuated rotating laser scanner. *Journal of Field Robotics*, 33(1), 103–132.

Kazhdan, M., & Hoppe, H. (2013). Screened poisson surface reconstruction. *ACM Transactions on Graphics (ToG)*, 32(3), 29.

Kordić, B., Đapo, A., & Pribičević, B. (2012, May). Application of terrestrial laser scanning in the preservation of fortified caves. In: *FIG Working Week 2012: Knowing to Manage the Territory, Protect the Environment, Evaluate the Cultural Heritage.* https://www.researchgate.net/publication/265025785_Application_of_Terrestrial_Laser_Scanning_in_the_Preservation_of_Fortified_Caves

Kregar, K., Vrabec, M., & Grigillo, D. (2019, January). Developing a robust workflow for acquisition of high-resolution full-3D cave topography, surface topography integration, and digital structural mapping. In: *Geophysical Research Abstracts* (Vol. 21). Copernicus Publications.

Kuda, F., Kajzar, V., Divíšek, J., & Kukutsch, R. (2014). *Aplikace pozemního laserového skenování v geovědních disciplínách.* Praha (Ústav geoniky Akademie věd České republiky, v.v.i.).

Leonov, A.V., Anikushkin, M.N., Bobkov, A.E., Rys, I.V., Kozlikin, M.B., Shunkov, M.V., . . . & Baturin, Y.M. (2014). Development of a virtual 3D model of Denisova Cave in the Altai Mountains. *Archaeology, Ethnology and Anthropology of Eurasia*, 42(3), 14–20.

Lerma, J.L., Navarro, S., Cabrelles, M., & Villaverde, V. (2010). Terrestrial laser scanning and close range photogrammetry for 3D archaeological documentation: The Upper Palaeolithic Cave of Parpalló as a case study. *Journal of Archaeological Science*, 37(3), 499–507.

Lyons-Baral, J. (2012). Using terrestrial LiDAR to map and evaluate hazards of Coronado Cave, Coronado National Memorial, Cochise County, AZ. *Arizona Geology Magazine*, Summer, 1–4.

Marsico, A., Infante, M., Iurilli, V., & Capolongo, D. (2015). Terrestrial laser scanning for 3D cave reconstruction: Support for geomorphological analyses and geoheritage enjoyment and use. In *Hydrogeological and Environmental Investigations in Karst Systems.* Springer, 543–550.

McFarlane, D.A., Buchroithner, M., Lundberg, J., Petters, C., Roberts, W., & Van Rentergem G. (2013). Integrated three-dimensional laser scanning and autonomous drone surface-photogrammetry at Gomantong caves, Sabah, Malaysia. In: Bosak, P., & Filippi, M. (Eds.), *Proceedings of the 16th International Congress of Speleology.* Brno, 2, 317–319. *Volume 2 International Congress of Speleology ICS Proceedings.* KIP Talks and Conferences. 13. https://digitalcommons.usf.edu/kip_talks/13

Milius, J., & Petters, C. (2012). Eisriesenwelt—from laser scanning to photo—realistic 3D model of the biggest ice cave on Earth. In: Jekel, T., Car, A., Strobl, J., & Griesebner, G. (Eds.), *GI-Forum 2012: Geovisualization, Society and Learning.* WichmannVerlag, Heidelberg: Salzburg, Austria, 513–523.

Mohammed Oludare, I., & Pradhan, B. (2016). A decade of modern cave surveying with terrestrial laser scanning: A review of sensors, method and application development. *International Journal of Speleology*, 45, 71–88.

Murphy, P.J., Parr, A., Strange, K., Hunter, G., Allshorn, S., Halliwell, R.A., . . . & Westerman, A.R. (2005). Investigating the nature and origins of Gaping Gill Main Chamber, North Yorkshire, UK, using ground penetrating radar and lidar. *Cave and Karst Science*, 32(1), 25.

Nash, G.H., & Beardlsey, A. (2013). The survey of Cathole Cave, Gower Peninsula, South Wales. *Proceedings of the University of Bristol Spelaeological Society*, 26(1), 73–83.

Nocerino, E., Menna, F., Farella, E., & Remondino, F. (2019). 3D virtualization of an underground semi-submerged cave system. *International Archives of the Photogrammetry, Remote Sensing and Spatial Information Sciences* (2/W15), 857–864.

Novoveský, A. (1975). *Technická správa, Domica 111-I-13.* Geologický prieskum, n.p., Geologická služba podniku, geologická oblasť Rožňava. Slovenské múzeum ochrany prírody a jaskyniarstva.

Núñez, M.A., Buill, F., & Edo, M. (2013). 3D model of the Can Sadurní cave. *Journal of Archaeological Science*, 40(12), 4420–4428.

Perperidoy, D.G., Tzortzioti, E., & Sigizis, K. (2010). A new methodology for surveying and exploring complex environments using 3D scanning. In: *FIG Congress*, 1–14. https://fig.net/fig2010/

Peterson, T., & Berg, J. (2001). *Karst mapping with geophysics at Mystery Cave State Park, Minnesota*. Minnesota Department of Natural Resources Ground Water and Climatology Section Report, 10.

Petrović, A.S., Ćalić, J., Spalević, A., & Pantić, M. (2018). Relations between surface and underground karst forms inferred from terrestrial laser scanning. *Geological Society, London, Special Publications*, 466(1), 107–120.

Petters, C., Milius, J., & Buchroithner, M.F. (2011). Eisriesenwelt: Terrestrial laser scanning and 3D visualisation of the largest ice cave on Earth. In: *Proceedings of the European LiDAR Mapping Forum*. Salzburg, Austria, 10 p. IEEE.

Plan, L., Roncat, A., & Marx, G. (2013). Detailed morphologic analysis of palaeotraun gallery using a terrestrial laser scan (Dachstein-Mammuthöhle, upper Austria). *Proceedings of the 16th International Congress of Speleology*, 1, 399–401.

Potenziani, M., Callieri, M., Dellepiane, M., Corsini, M., Ponchio, F., & Scopigno, R. (2015). 3DHOP: 3D heritage online presenter. *Computers & Graphics*, 52, 129–141.

Potree. (2021). www.potree.org/ (accessed on 20 August 2021).

Pucci, B., & Marambio Castillo, A.E. (2009). Olerdola's cave, Catalonia past and present: A virtual reality reconstruction from terrestrial laser scanner and gis data. In *3rd International Workshop 3D Virtual Reconstruction and Visualization of Complex Architectures*, 1–14. ISPRS.

Pukanska, K., Bartoš, K., Bella, P., & Sabová, J. (2017). Comparison of non-contact surveying technologies for modelling underground morphological structures. *Acta Montanistica Slovaca*, 22(3), 246–256.

Radicioni, F., Rossi, G., Tosi, G., & Marsili, R. (2019, May). Non contact shape and dimension measurements by LIDAR techniques of one of the biggest Italian caverns. *Journal of Physics: Conference Series*, 1249(1), 012019. IOP Publishing.

Rapidlasso GmbH. LAStools. https://rapidlasso.com/lastools/ (accessed on 1 August 2019).

Robson-Brown, K., Chalmers, A., Saigol, T., Green, C., & D'errico, F. (2001). An automated laser scan survey of the Upper Palaeolithic rock shelter of Cap Blanc. *Journal of Archaeological Science*, 28(3), 283–289.

Roncat, A., Dublyansky, Y., Spötl, C., Dorninger, P., & Pfeifer, N. (2011). A full-3D laser-scan mapping of a hypogene cave: A morphogenetic study of Märchenhöhle, Austria. *Geophysical Research Abstracts. EGU 2011, Vienna*, 13, 14039.

Rüther, H., Chazan, M., Schroeder, R., Neeser, R., Held, C., Walker, S.J., . . . & Horwitz, L.K. (2009). Laser scanning for conservation and research of African cultural heritage sites: The case study of Wonderwerk Cave, South Africa. *Journal of Archaeological Science*, 36(9), 1847–1856.

Santagata, T., Lugli, S., Camorani, M.E., & Ercolani, M. (2015). Laser scanner survey and tru view applications of the "Grotta della lucerna" (Ravenna, Italy), a roman mine for lapis specularis. In *Hypogea 2015: Proceedings of International Congress of Speleology in Artificial Cavities: Italy*, Rome, 11/17 March 2015, 1, 411–416.

Santos Delgado, G., Martínez Rubio, J., Silva Barroso, P.G., Sánchez Moral, S., Cañaveras Jiménez, J.C., & De la Rasilla Vives, M. (2012). Contribución al conocimiento de la cueva de El Sidrón (Piloña, Asturias) con técnicas de láser escáner 3D. In: González, A., et al. (Eds.), *Avances de la Geomorfología en España 2010–2012. Actas de la XII Reunión Nacional de Geomorfología, Santander*, 17–20 September 2012, 255–258. Editorial Universidad de Cantabria, Santander.

Scheiblauer, C. (2014). *Interactions with Gigantic Point Clouds*. Ph.D. Thesis, Vienna University of Technology, Vienna, Austria, 203.

Schütz, M. (2016). *Potree: Rendering Large Point Clouds in Web Browsers*. Engineer Diploma Thesis, Vienna University of Technology, Vienna, Austria, 92.

Schütz, M., Ohrhallinger, S., & Wimmer, M. (2020). Fast out-of-core octree generation for massive point clouds. *Computer Graphics Forum*, 39(7), 1–13.

Scopigno, R., Callieri, M., Delleppiane, M., Ponchio, F., & Potenziani, M. (2017). Delivering and using 3D models on the web: Are we ready? *Virtual Archaeology Review*, 8, 1–9.

Silvestre, I., Rodrigues, J.I., Figueiredo, M., & Veiga-Pires, C. (2015). High-resolution digital 3D models of Algar do Penico Chamber: Limitations, challenges, and potential. *International Journal of Speleology*, 44(1), 25–35. http://doi.org/10.5038/1827-806X.44.1.3

Stipanov, M., Bakarić, V., & Eškinja, Z. (2008). ROV use for cave mapping and modeling. *IFAC Proceedings Volumes*, 41(1), 208–211.

Šupinský, J., Kaňuk, J., Hochmuth, Z., & Gallay, M. (2019). Detecting dynamics of cave floor ice with selective cloud-to-cloud approach. *The Cryosphere*, 13(11), 2835–2851.

Thibault, G. (2001). 3D modeling of the Cosquer cave by laser survey. *International Newsletter on Rock Art, No. 28*, 25–29. https://www.bradshawfoundation.com/inora.php

Triantafyllou, A., Watlet, A., Le Mouélic, S., Camelbeeck, T., Civet, F., Kaufmann, O., . . . & Vandycke, S. (2019). 3-D digital outcrop model for analysis of brittle deformation and lithological mapping (Lorette cave, Belgium). *Journal of Structural Geology*, 120, 55–66.

Tsakiri, M., Sigizis, K., Billiris, H., & Dogouris, S. (2007, September). 3D laser scanning for the documentation of cave environments. In *11th ACUUS Conference: Underground Space: Expanding the Frontiers*. http://www.minetech.metal.ntua.gr/

Tyszkowski, S., Kramkowski, M., Wisniewska, D., & Urban, J. (2016, April). Use of terrestrial laser scanning for the documentation of quaternary caves. In *EGU General Assembly Conference Abstracts* (Vol. 18). Copernicus Publications.

Ullrich, A., Schwarz, R., & Kager, H. (2003). Using hybrid multi-station adjustment for an integrated camera laser-scanner system. https://publik.tuwien.ac.at/files/PubDat_119447.pdf

Wenger, R. (2004). La balise de positionnement U-GPS (Underground-GPS). *ISSKA Rapport Annuel* (Swiss Institute for Speleology and Karst Studies, La Chaux-de-Fonds 2004), 13–14. http://www.isska.ch/pdf/Fr/Rapport_annuel/rapp_annuel_04.pdf

Westerman, A.R., Pringle, J.K., & Hunter, G. (2003). Preliminary LIDAR survey results from Peak Cavern Vestibule, Derbyshire, UK. *Cave and Karst Science*, 30(3), 129–130.

Yakar, M., Ulvi, A., & Toprak, A.S. (2016). The use of laser scanner in caves, encountered problems and solution suggestion. *Universal Journal of Geoscience*, 4(4), 81–88.

Zeid, N.A., Bignardi, S., Russo, P., & Peresani, M. (2019). Deep in a Paleolithic archive: Integrated geophysical investigations and laser-scanner reconstruction at Fumane Cave, Italy. *Journal of Archaeological Science: Reports*, 27, 101976.

Zlot, R., & Bosse, M. (2014). Three-dimensional mobile mapping of caves. *Journal of Cave & Karst Studies*, 76(3).

5 Digitizing Giant Skeletons with Handheld Scanning Technology for Research, Digital Reconstruction, and 3D Printing

Jesse Pruitt, Tim Gomes, Evelyn Vollmer, and Leif Tapanila

HARDWARE AND SOFTWARE

Three different handheld scanners were used for digitizing the blue whale skeleton and orca skull: FARO Design S Arm™(FDSA), FARO Edge Arm™ (FEA), and a Creaform Go!SCAN 50™ (CGS50). The FARO Arm scanners are laser-based scanning systems that use a visible green (FDSA)/red (FEA) laser and a camera to capture 3D surface data. Both arm scanners require the use of a heavy-duty tripod and ideally an unobstructed area of 8 feet in diameter to allow for 360° free movement of the articulated arms and access to a power source. FARO uses a system called the laser line probe (LLP) for non-contact surface digitization. The LLP is encoded to the arm itself and is able to accurately relay the laser positioning data to the software. The attached camera records the laser as it deforms on the surface of the object being scanned, and this is then used to generate point cloud data to an accuracy of 18 microns (0.0007 inches). The cameras positioned on each side of the projector that are mounted at a slight angle to allow for accurate triangulation of the projected pattern on the surface and an RGB camera to capture color information. The cameras look for and record the deformation of the projected pattern to accurately build a surfaced mesh. It is considerably more mobile and only requires a laptop and a power source. The CGS50 is capable of a maximum resolution of 0.5 mm with multiple scan passes and an accuracy 0.3 mm (0.004 inches). Figure 5.1 shows a small selection of scanner setups and locations used to capture the blue whale and orca in 3D.

DOI: 10.1201/9780429327575-5

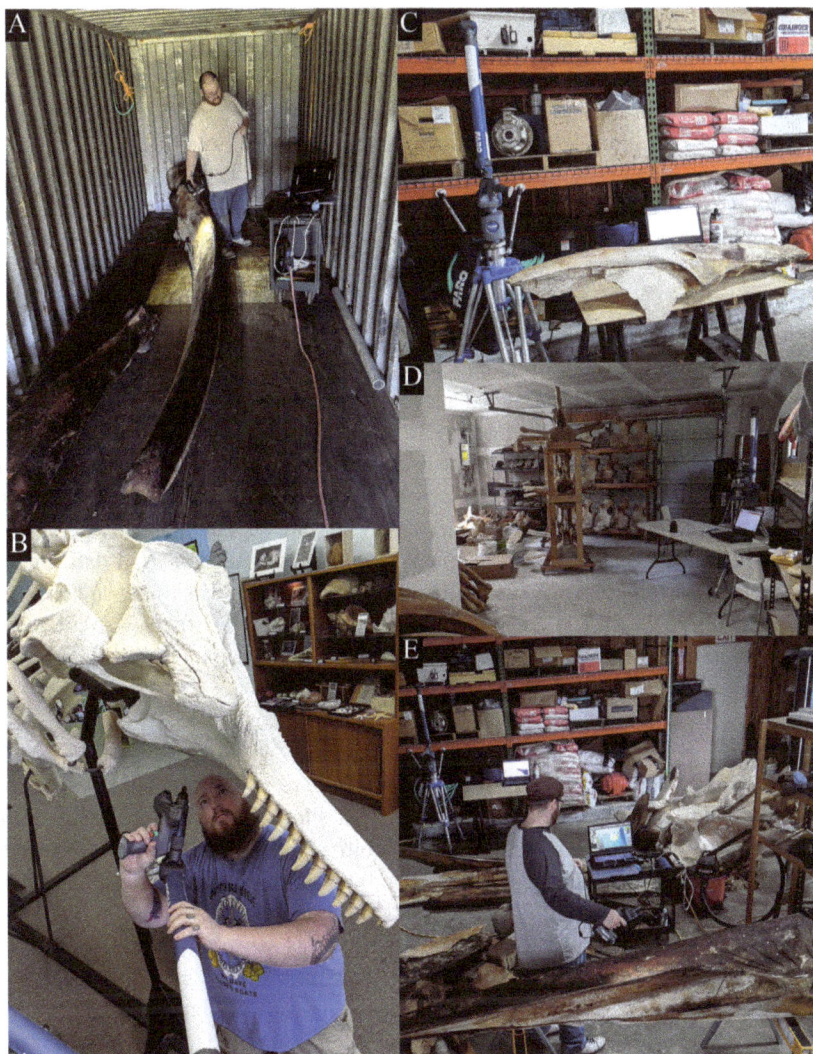

FIGURE 5.1 Various scanner setups and locations for scanning the blue whale and orca. (A) Gomes scanning the lower jaw of the blue whale using the CGS50. (B) Pruitt scanning the ventral surface of the orca cranium using the FDSA. The specimen is mounted and displayed at the NOYO Center for Marine Science in Fort Bragg, CA. (C) Makeshift workbench for scanning the blue whale cranium pieces with the FDSA, using the magnetic stabilizer arms on the shelving in the background. (D) Both FARO scan arms set up in the temporary housing garage, using collapsible tables as workbenches. (E) Gomes scanning the maxilla of the blue whale using the CGS50 and sawhorses as a platform.

For remote scanning, high-powered gaming workstation laptops are recommended to capture scan data. This project used MSI Custom Build Series with i7–8700 processors, Nvidia Quadro P3200, 64 GB of DDR4 RAM, and solid-state hard drives. These laptops allow for fast and lag-free data capture with high-resolution scan settings on large objects. External 2 TB hard drives were used for data backup at the end of each day.

Data capture for both FARO systems used InnovMetric Polyworks™, while Creaform VXElements™ was used for the CGS50. Polyworks has a modular software system and remote licenses, making it easy to work in isolated areas for extended periods of time without access to the Internet as licenses can be checked out on each laptop for up to 90 days. The Polyworks Align module was used for data capture and data alignment and for adjusting the scanning parameters and calibrations of the FARO Arm systems. VXElements is proprietary software specific to the Creaform lineup of scanners and is an all-in-one package that handles scan settings, machine calibration, data alignment, and scan surfacing. This software uses a down-loaded license key and does not need access to the Internet to verify the license after the initial setup making it ideal for remote scanning work.

Merge and Edit modules of Polyworks were used to generate a surface and export .obj files for data captured with the FARO Arm scanners, and Geomagic Wrap was used for data editing and cleanup. VXElements automatically generates a mesh surface as it scans, but Geomagic Wrap was also used for final data cleanup after it was exported from VXElements. Pixologic ZBrush was then used as a final step to optimize mesh topology, clean scan errors and artifacts, articulate and merge skeletal elements, and prepare a manifold (watertight) mesh for 3D printing.

SCANNER SETTINGS AND CALIBRATION

The FARO Arm scanner settings for this project were a compromise between high resolution and data management. Ideally, data should be captured at as high a resolution as possible without making it prohibitively difficult or time-consuming to process. The following settings were used on both of the FARO Arm scanners: max point-to-point distance of 0.25 mm, a standard deviation of 0.15 mm, max angle of 75°, max edge length of 0.75 mm, laser exposure was set to automatic normal, and data capture at 1:1. The calibration process for both the Design S and Edge Arms is the same and is conducted each time the arms are moved to a new location/setup. Using FARO's plugins for PolyWorks, the scanner is calibrated using a manufacturer-supplied measurement plate to achieve plane compensation. Then the rotational axes are calibrated using ballpoint compensation. After these steps are completed, the resulting calibration file is stored in PolyWorks and remains valid until the scanner is moved.

The default scanning resolution for the CGS50 is 2.00 mm, but through the supplied software, the final data can be exported at 0.5 mm. The system initially captures a surface at a resolution of 0.5 mm, but then downsamples it to 2 mm to prevent the system from becoming overwhelmed by huge amounts of data and lagging on larger scans. There are a few options to choose from for scan settings to aid in the alignment of the scan data: surface features, surface features and texture, surface features and target, or just targets. Surface features were selected for most of this project as the bones had sufficient surface detail to generate their own control points. However, on bones that had large areas of featureless surfaces, a hybrid surface feature/target-based approach was taken. The reference targets used in this process were vendor-supplied 1 cm self-adhesive, reflective circles that the system is designed to detect and track. These are placed randomly along a surface, ensuring that at least 3 of them are within the 15 cm x 15 cm field of view at any one time. Use of these targets was kept to a minimum as they result in 1 cm circles of missing data that show up in the final mesh as a smooth, featureless area. Creaform includes a calibration plate with the system that should be used before every scan. This plate is a surface with a known pattern of targets that ensures the cameras are tracking properly. It is recommended to calibrate between successive large scans as the data can start to drift if the scanner is used for long periods. The software also allows for the calibration of the light intensity of the projected grid. This is very useful as using the automatic setting can lead to tracking issues resulting in scan errors.

SCANNING AND DATA PROCESSING

Scanning the very large skull of the blue whale posed some unique challenges due to the size and weight of the elements involved. The mandibles alone were approximately 11 feet (3.35 m) long and, because they were still full of grease from when the animal was alive, weighed more than 500 pounds (227 Kg) each. The largest section of the cranium was stored on a reinforced pallet in a near-vertical orientation resting on the occipital region, making it much easier to move into a favorable scanning location using a pallet jack. However, it had to be repositioned so that a scan of the occipital area could be taken to complete the data set. This was done by hand, with four people assisting in tipping it forward and gently lowering it onto stacks of high-density furniture foam to ensure the safety and stability of the specimen. The remaining elements of the skull, while still large and heavy, were relatively easy to move and position by two people.

The largest section of the cranium was scanned in five separate groups using the CGS50 with a 20–30% overlap between each group to ensure that everything would join cleanly during the final alignment and merging process. Initial trials using the natural features of the bone to maintain tracking failed due to the relative smoothness of the surface. Therefore, tracking targets were placed on the surface as sparsely as possible and placed in areas devoid of details to minimize data loss. After each scan group was completed, the targets were moved to the next area with the exception of the targets along the area of overlap, which were left in place to assist with subsequent alignment. The left and right maxilla, the largest intact pieces of the anterior section of the skull, were positioned horizontally on a pair of sawhorses. This placed them at an optimal height for scanning comfortably. They were scanned in four groups, using tracking targets because grease present in the bone made it darker, making tracking via features very problematic. The mandibles were scanned on the floor of the Conex shipping container they were stored in, using a similar process as the maxilla with four scan groups, with each group captured using a combination of tracking targets and natural features. Scans were taken as close as possible to the floor of the container to ensure there would be enough overlap for joining once the mandible was rolled over.

3D scanning with the FARO Arms was carried out in a similar fashion minus track-ing targets. A makeshift table was created using a pair of sawhorses and an 8 x 4 foot sheet of plywood. The Design S arm was mounted to the tripod at the back left corner of this table. Starting with the smaller fragments of the cranium, each piece was placed on the table, and the arm was put through a range-of-motion test to confirm that it could be scanned from all angles without overextending and stressing the arm joints. The smaller pieces were scanned in two groups, an A and a B scan. As the pieces became larger, it was necessary to increase the number of scan groups to prevent the laptop from being overwhelmed with too much data and fail during post-processing. The larg-est piece, a >12 feet (3.65 m) long premaxilla, was scanned in four groups.

All of the post-cranial skeleton elements were scanned using the FDSA and the FEA, with two scan groups per bone. The FDSA has a much higher rate of cap-ture and a longer arm than the FEA, so it was used to scan the larger pieces such as the thoracic vertebrae, ribs, and the large arm bones (humerus, ulna, and radius), while the FEA was used on the smaller lumbar and caudal vertebrae and flipper bones. As the blue whale was not fully matured, many of the epiphyseal plates on the vertebrae had not fully fused and separated from the main vertebral bodies, and several of the neural spines on the thoracic vertebrae had broken when the animal was beached. All of these pieces were scanned separately and would be positioned when the skeleton was digitally articulated. Figure 5.2 shows the FDSA scan data collected from vertebrae 13. There was also three five-gallon (circa 20L) buckets of fragmentary bone pieces that we scanned with the hope of being able to locate them as the skeleton was digitally reconstructed.

The orca skeleton is mounted and on display in the main office of the NOYO Center for Marine Science, CA. The cranium of the orca was scanned in place using the FDSA and the CGS50 to ensure that as much of the internal nasal cavity as possible was captured, as the CGS50 can capture data in deep cavities much easier than the laser-based FARO system (Figure 5.3). The left and right mandibles were scanned separately with the FDSA.

As described previously, the CGS50 generates a 3D mesh surface in real time instead of a point cloud like most other 3D scanners. This means that after a scan is complete, there is no alignment to do within that scan group, but you can select indi-vidual photo frames and delete them if there are areas of poor data. To align all of the scan groups for the cranium, they were initially opened in one workspace and any superfluous data, including the surface of the pallet the cranium was resting on, were deleted from each scan group. The merged scan is then selected with the surface best-fit alignment option. Scan one is set as a fixed scan, and using a pre-alignment manual mode, up to 11 points on each surface to be joined are selected. After the points have been matched between the fixed and mobile scans, best fit is selected and the software will perform a global registration between the two scans. If success-ful, the alignment is accepted and the next mobile scan data set is selected—these steps are repeated until all scans have been aligned. Finally, the scans are merged. This function deletes any overlap and generates a single mesh from all the scans. It is recommended to accept the 'keep original scan' option in case any errors are introduced during the merge. After reviewing the merged surface, the resolution is changed to 0.5 mm from 2.00 mm and the model is exported as an .obj file. The file

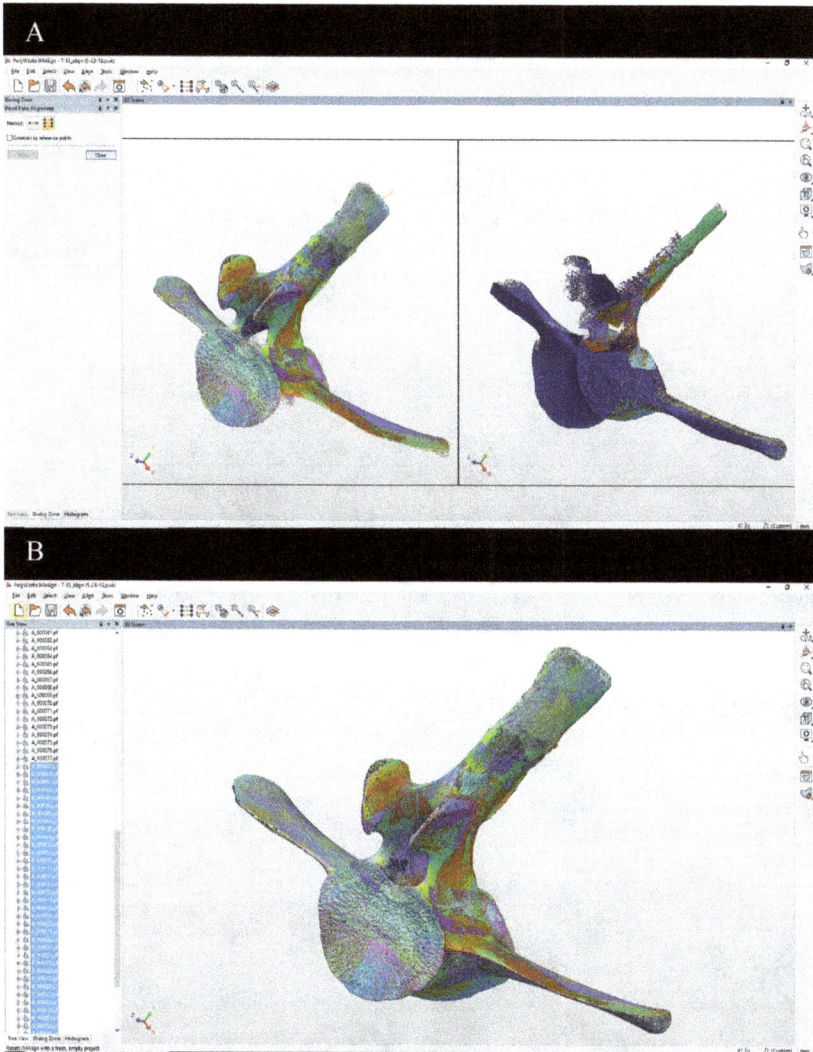

FIGURE 5.2 FDSA scan data of the blue whale thoracic vertebrae number 13. Each color patch on the point cloud represents an individual scan pass with the arm, 125 total scan passes for this vertebrae. (A) The anterior and posterior scans of the vertebrae are being aligned using the Points Pair alignment tool in the PolyWorks Align module. Overlapping details are selected on each set of scan data, and the software will complete a rough alignment after enough points have been selected. (B) The fully aligned set of data after completing the Best Fit, or Global Alignment process inside of PolyWorks Align.

is then opened in Geomagic Wrap for post-processing during which it is converted to points so that outliers can be filtered out. It is resurfaced using a wrap function. Any large holes in the mesh are manually edited and closed using a curve fill function, and finally, the 'mesh doctor' operation in this software is run to remove spikes,

A

B

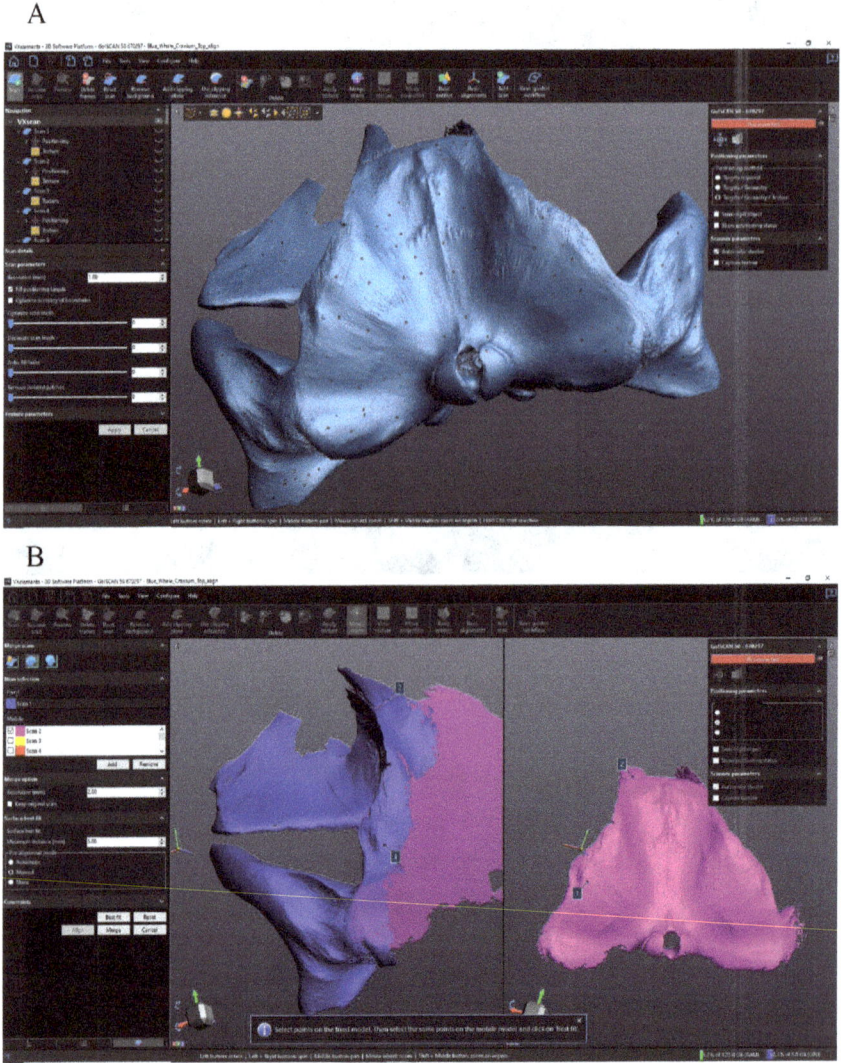

FIGURE 5.3 The top half of the blue whale cranium data scanned with the CGS50. This is a surfaced but non-manifold mesh as opposed to a standard point cloud. (A) This is the aligned and merged mesh of the top of the blue whale cranium after manually aligning and merging seven individual scans. The white dots seen on the surface are the tracking targets placed on the surface to ensure clean scan data across such a large surface area. (B) Combining scans one and two using the manual best fit alignment options in the merge scans tool. Similar details on the surface are selected on each scan in the overlapping areas then a best fit alignment is ran between them. After the data is perfectly aligned, the next scan is selected and aligned to the first two and the process is repeated until all of the scans have been aligned. This new set of data is saved as a separate file to be later merged with the bottom half of the cranium's scan data.

close any small holes, and check for overlapping data. This 'cleaned' file is then decimated to 99.9% to eliminate any unused floating points, moved to the origin, and again exported as an .obj file. In most cases, this file would be ready to use for 3D printing or morphometric analysis, but since this is only a portion of the cranium, it will be used as the base for reassembling the skull. This same workflow was used on the remaining scans from the CGS50.

The FARO Arm 3D scanners generate point cloud data as they scan, and this data is saved as an Align file for each bone scanned, with each file having a minimum of two scan orientations saved as a group with a letter designation such as A, B, or C. In the Align module of the manufacturer's software, all of the scans from one group are selected and the data is aligned using a best-fit alignment tool with the following settings—tolerance of 0.5 mm, maximum distance of 2.00 mm (this is the maximum distance a single scan file can move in each alignment iteration), subsampling of 1/25 (to speed up the process), automatic histogram range, refresh mode of every 4 iterations, unlimited number of iterations, and a convergence tolerance of 0.000. This alignment is allowed to run until a convergence has been reached or the scans stop moving visually. This is an approximate alignment that will be refined later in the process. After all of the scan groups have been individually aligned, they are then aligned to each other using a points pair alignment tool. As before, the first scan group is used as the fixed scan, and the subsequent groups are aligned to it one at a time by selecting common surface features manually via control points. A minimum of eight control points per scan was used. The next scan group is then selected, and the process is repeated until all of the data has been aligned. At this point, a 'best-fit alignment' tool is used to refine the global registration of all of the data. Once complete, the data is saved as a new file, maintaining the original scan data as an archival copy. This data is still in a point cloud format and must be meshed to produce a surface using the Merge module in PolyWorks before it can be ready for additional use. In the Merge module, the smoothing level is set to none, max distance 2.000, surface sampling step 0.10, a standard deviation of 0.08, and subdivision is changed to # merging jobs to prevent automatic downsampling of the data. After the merge process is completed, the resulting surfaced mesh is automatically opened in the Edit module and exported as an .obj file. This .obj file is opened in Geomagic Wrap and edited in the same workflow for the CGS50 data described before. The only exception to this workflow was the epiphyseal caps from the vertebrae where only the external surface was scanned. These were left non-manifold, with the back open, as they would be merged with their respective vertebrae in the next editing process.

All the models were then imported into ZBrush for final cleanup, articulation, and optimizing the topology for web distribution. To make the data easier to work with at this stage, all of the individual elements were simplified so that the entire skeleton can be opened and articulated while not overwhelming the software or the computer hardware. By putting each bone through a process called remeshing in ZBrush, a low-resolution mesh was created that is approximately 5–10% of the total polygon count of the original bone. If that low-resolution mesh is subdivided, you can project the details from the original high-resolution mesh onto it and will end up with a mesh that can be reduced in polygon count for articulation but bumped back up to a high-resolution mesh for final export and renders. This process, illustrated in Figure 5.4, is as follows:

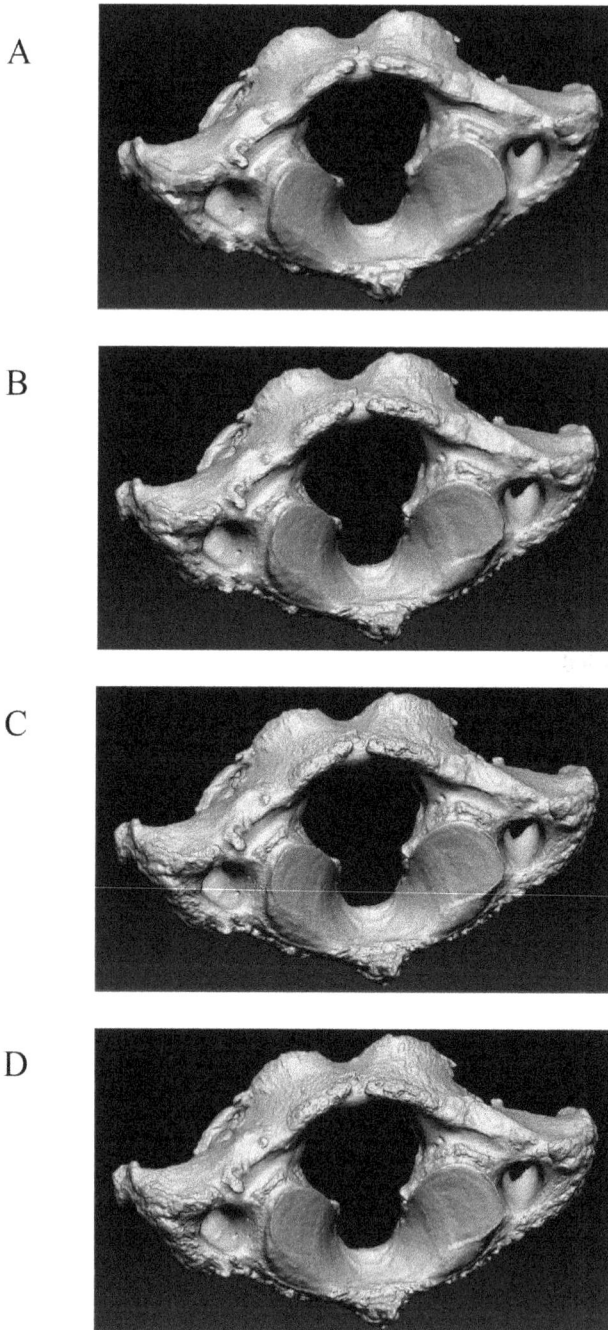

FIGURE 5.4 Blue whale atlas vertebrae displayed at multiple subdivision levels after the details from the original high-resolution mesh have been projected onto it. (A) Subdivision 1 (B) Subdivision 2 (C) Subdivision 3 (D) Subdivision 4.

each bone is duplicated and the copy is sub-sampled to approximately 5,000 polygons using the Remesh function. This produces a set of clean quadrangle polygons in place of the scanned triangle polygons. The details are then projected from the high-resolution copy onto the low-resolution mesh using the Project All function. This will run an algorithm that will try to match the features of the low-resolution mesh to the high-resolution mesh but at the lowest subdivision level—not much will happen visually. The mesh is subdivided and the projection process repeated until the retopologized mesh has all of the detail of the original. Each step up in subdivision levels will produce a much more detailed mesh on the lower resolution copy, usually four to five division levels is enough to ensure no details are lost. The high-resolution mesh is deleted after the projection process has been completed, and you are left with a mesh that has all the detail but now has multiple subdivision levels so it can be dropped down to approximately 5,000 polygons. This will use fewer computer resources as the bones are moved into position and articulated, and with all of the meshes at their lowest subdivision level, the total file size for the articulation file will be much lower than if it had only high-resolution meshes, usually a 4–6 Gb size difference. This process of remeshing and projection was done for every bone in the data set.

The skull was digitally restored first as it was the foundation that the rest of the 3D skeleton model would be built upon. The pieces were matched together like a giant bone puzzle and aligned as best as possible. When as many pieces as possible were assembled, the individual pieces were set to the highest subdivision level and then merged into a single object. This was the 3D mesh that was used as the final output for the skull and uploaded to www.morphosource.org. This workflow was repeated for any of the broken vertebrae as well as adding in the epiphyseal plates. The vertebrae were articulated in groups to keep the data size manageable. These groups were the cervical, thoracic, lumbar, and caudal. The flipper and ribs were also assembled as their own groups. After all of these groups had been articulated and the positioning of the elements more or less optimized, they were brought into the same data set as the skull and positioned together to complete the 3D skeletal articulation. The final positioning of all the bones, shown in Figure 5.5, was refined as a whole to bring the skeleton into a lifelike position and be aesthetically pleasing. The skeletal assembly was exported as a single file to be uploaded, albeit in a decimated state as the full-resolution skeleton would be too large to be opened on most computers. Each bone was also rendered on a standard background and exported separately to be uploaded to the NOYO Blue Whale #1817597 project on www.morphosource.org. All of this data is available to be downloaded for free and is open access and licensed as public domain. The articulated skeleton can also be viewed online at www.sketchfab.com by searching for NOYO Blue Whale Skeleton.

3D PRINTING

To prepare the orca skull for full-scale 3D printing, ZBrush was used to clean any remaining scanning artifacts and optimize the surface topology. In general, 3D slicing software does not care about the order of the polygons of a model, but it can be sensitive to the number of polygons. Initial trials showed that a model with a maximum of 3.2 million polygons was ideal for most 3D printing slicing programs and produced G-code that was free from errors that did not overwhelm the 3D printer's onboard software. The

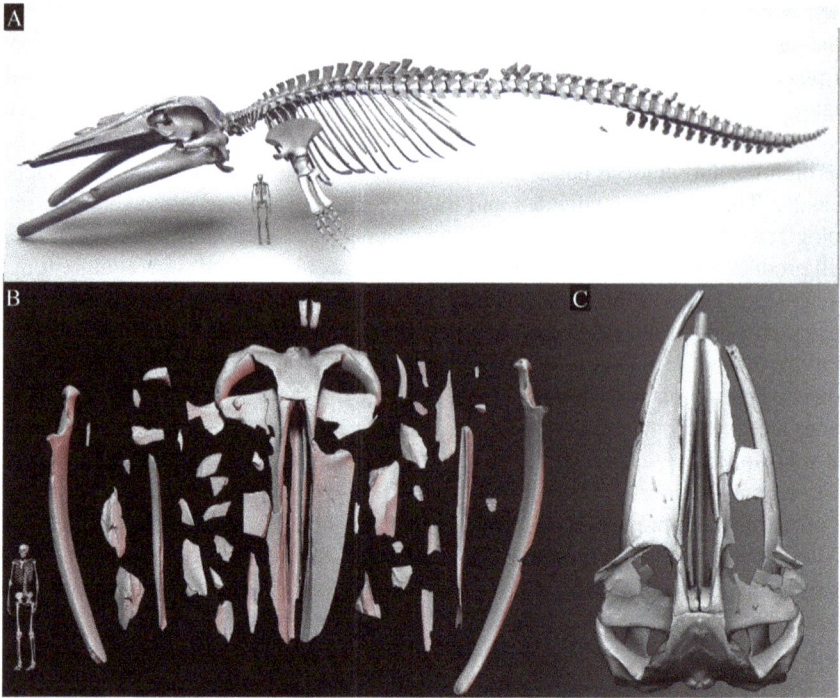

FIGURE 5.5 Final articulation and renders of the blue whale skeleton. (A) The full skeleton is rendered here with a 1.8 m-tall human skeleton for size reference. The damage on the skull is the result of the animal becoming lodged in rocks postmortem as it was beached. The damaged thoracic vertebrae are the results of the ship strike that is suspected to have resulted in the animal's death. (B) All of the individual pieces of the skull that were digitally reassembled. Much of the cranium had become too badly damaged to be identifiable and thus was not scanned and included here. (C) The final assembly of the blue whale skull.

physical size of the 1:1 scale skull would not fit on the largest 3D printer in the IVL lab in one piece, so it had to be subdivided into 17 separate parts, as shown in Figure 5.6. These parts were then 3D printed and glued, with epoxy, back together to complete the life-size replica. This process did result in a much higher resolution final 3D printed model since each part could be kept below the 3.2 million polygon limit.

To split the 3D model, a digital box was created that represented the build volume dimensions of the 3D printers in the IVL so that the size of each 3D printed part could be maximized and, therefore, minimize the total number of separate parts. Consideration was also given to the shape of the parts with a priority on avoiding creating any thin edges or small features that would be difficult to print or require a large amount of support material during printing. Using the build volume box as a guide, simplified 3D planes were used to perform Boolean cut operations in ZBrush. This splitting process resulted in 17 parts that could fit within the 400 x 400 mm build plates of the 3D printers. After splitting, if the models exceeded the 3.2 million polygon limit, the polygon count was reduced using the

Decimation Master plugin, setting the exact number of polygons desired. The split files were then exported in STL format and named with the general location (e.g., Left_premaxilla, Right_occipital).

Each model was then opened in PrusaSlicer to orient it optimally on the build plate and check the mesh for errors using the included Netfabb repair function. The slicing software Simplify 3D was used to set the printing parameters. Each part was printed in polylactic acid filament (PLA) using a 0.4 mm nozzle, a 0.24 mm layer height, with 3 perimeters, 5 bottom layers, 6 top layers, 3% infill, 65 °C bed temperature, and 208 °C nozzle temperature. Printing supports were manually placed using Simplify 3D to optimize print speed and reduce material waste. The model was then sliced, checked layer by layer for any errors and potential printing issues, and exported as a G-code file specific to one of the three machines used for this project.

Post-processing of the 3D prints consisted of light sanding to remove any severe layer lines. The mating surfaces of each print were also sanded to increase bonding, and an epoxy resin was applied to fix any large defects. Post-processing time could be greatly reduced by decreasing layer height and slowing the print speed, but this would have taken considerably more time to print. The entire skull was then coated in two thin layers of resin to improve the surface finish and enhance the overall strength of the final piece. The resin took 24 hours to cure and was self-leveling, so it was best to work in thin layers to prevent pooling in cavities and reduce the appearance of fine detail on the print. The skull was lightly sanded to give the paint primer a surface to bond to, painted with a white base

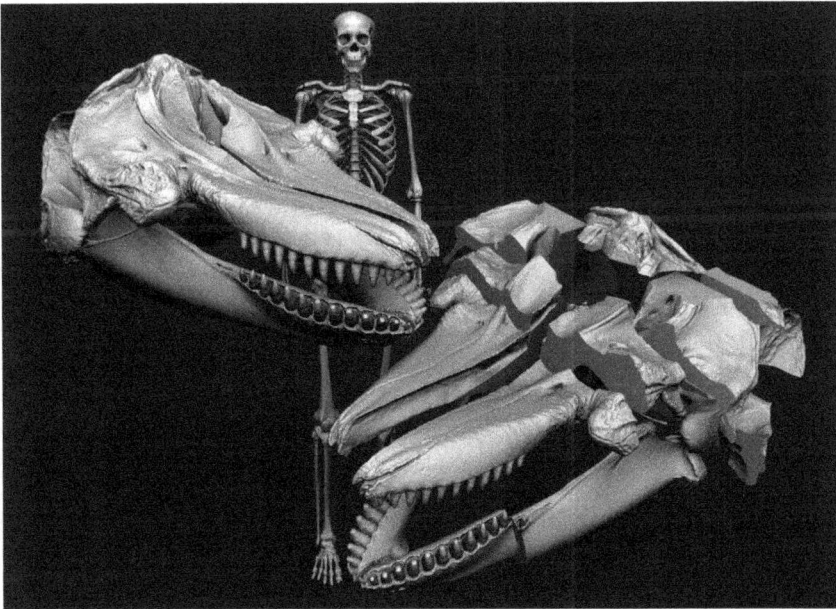

FIGURE 5.6 Final edited orca skull and the multiple part split file for 3D printing, 1.8 m tall skeleton for size reference.

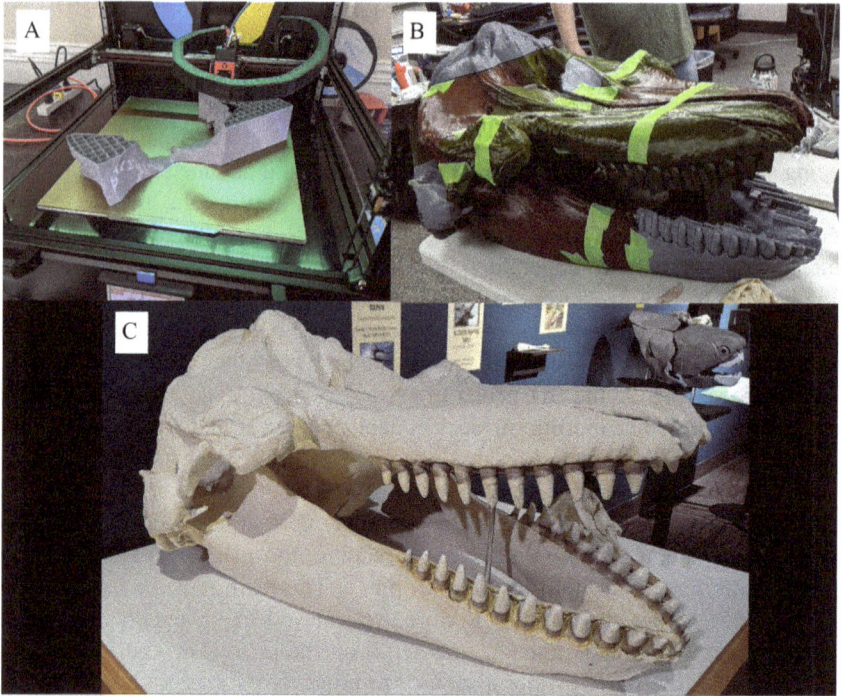

FIGURE 5.7   3D printing process of the orca skull. (A) Printing one of the pieces of the orca cranium on a Vivedino Troodon 400 corexy 3D printer, using PLA filament. (B) Orca skull prints have been assembled and epoxied in place. Painter's tape is used to help hold the pieces in place as the epoxy had not fully cured, but we were impatient and wanted a photo. (C) The final display of the painted orca skull in the Skulls Exhibit at the Idaho Museum of Natural History.

coat, and then coated with a dark wash layer to bring out the recessed details. Pastel pigment powders were used to add shading to the tooth roots and worn areas. The final step involved sealing the surface with a satin clear coat applied in several layers over a period of three days before putting the skull on display. This procedure is shown in Figure 5.7.

ACKNOWLEDGMENTS

We would like to thank the National Science Foundation (NSF) for funding the scanning project; Sheila Seimans at the NOYO Center for Marine Science for helping us with the logistics, planning, and handling of the specimens; Jeff Jacobson (NOYO) for his anatomical expertise and helping connect me with the right people to make this project possible; the City of Fort Bragg for allowing us to use their facilities for scanning and helping move the largest pieces of the cranium; and the amazing team of Iadho State university student interns that helped us process the huge amount of data: Mary Munoff, Joseph Tyler, and Lathen O'Neill. The project was funded by the Polar Programs office of the National Science Foundation (NSF award #1817597 to PI Leif Tapanila).

6 Mapping, Monitoring, and Visualising Stone Decay in the Urban Environment
Worcester College and New College, Oxford

John Meneely

INTRODUCTION

Numerous historic structures worldwide are constructed of limestone that deteriorates over time, frequently requiring costly repair or replacement. This decay has been shown to increase in polluted environments as airborne contaminants add to the physical and chemical processes that cause damage to stonework, often leading to cavernous and rapid recession of the original stone surface (Brimblecombe & Grossi 2008, 2009; Searle & Mitchell 2006; Rodriguez-Navarro & Sebastian 1996).

Traditionally, quantification of surface change on in-situ building blocks and laboratory-based weathering experiments were small-scale and relied upon mechanical techniques that made contact with the surface they were measuring. These were often time-consuming, endangered damaging the surface being measured, and required a statistical interpolation among a limited number of points (Diaz-Andreu et al. 2006; Inkpen 2007).

As Pope et al. (2002) said, "The best recommendation for slowing rates of deterioration is to limit human contact, including scholars".

Overcoming these obstacles requires a rapid, non-contact method for monitoring surface change using a dense network of measurement points. It was in search of improvement in the speed and precision of surface analysis that this study trialed the use of two 3D laser-based surface scanning systems as a means of accurately and non-destructively monitoring the progressive decay of carbonate building stone in laboratory-based experiments and in the field.

Poor air quality has arguably been linked with the deterioration of many building materials, most notably stone. Sulphur dioxide (SO_2) in rainwater is often associated with the sulphation and dissolution of limestone on surfaces subject to exposure. In contrast, dry deposition of SO_2 on sheltered stone surfaces can result in the formation

DOI: 10.1201/9780429327575-6

of gypsum crusts that are turned black by the inclusion of airborne particulates. These black crusts can also form on exposed surfaces in highly polluted environments, where the rate of surface gypsum formation exceeds that of dissolution by incident rain and surface runoff. This results in the uniformly black buildings that once characterized many of our cities. Recent years have seen substantial decreases in atmospheric pollution across many urban areas, particularly in Europe and North America (Fenger 1999). This has, in turn, prompted an assumption that rates of material deterioration must have similarly decreased, that it is safe to clean stonework without the rapid reappearance of black crusts, and that less will have to be spent in the future on the costly conservation of culturally valuable structures—"Pollution controlled damage to durable building materials seems to be over" Brimblecombe (2008). However, the relationship between air pollution and building stone decay may be more complex than previously expected, especially on historic limestone facades exposed for many years to multiple, superimposed processes of deterioration. Long-term monitoring, 1980–2010, at St Paul's Cathedral, London, (Inkpen et al. 2012) has identified possible hysteresis effects, whereby lower air pollution has not delivered the expected reductions in surface recession from simple dose/response relationships calculated from periods of increasing pollution.

"Analyses of measurements and modelling studies show the SO_2 concentrations in most UK cities have decreased dramatically over the last 40 years. As a result, current concentrations do not provide an adequate picture of cumulative exposure of the built environment" (Eggleston 1992).

Furthermore, dependence on solution loss as the principal agent of stone decay neglects the largely physical and more aggressive deterioration associated with the internal crystallization of soluble salts (Cardell et al. 2008). Salts can derive from a number of sources either directly or indirectly from atmospheric pollution, possible inheritance from earlier episodes of poor air quality, and from the capillary rise of groundwater. Their efficacy as instruments of decay is well recorded, and this, in turn, is controlled by the complicated moisture and thermal systems experienced on building facades that result from their complex geometry and surface morphologies, which are themselves constantly undergoing change as deterioration creates its own surface relief (Cardell et al. 2008; Turkington et al. 2003; Sass & Viles 2010; Smith et al. 2011).

It is because of these complications that air quality improvements may, as this study will demonstrate from monitoring historic limestone walls in Oxford, UK, result in a counter-intuitive acceleration of deterioration.

Oxford is typical of many historic cities in that it contains a diverse range of buildings and monuments of great cultural value. Its built heritage is predominantly constructed of Jurassic age oolitic limestone and spans the period from the late 11th to the early 21st century. Oxford's air pollution history is well-documented and shows comparable trends to that of other major UK cities with a clear non-linear decline of around 75% in smoke and particulate concentrations over the last 50 years (from > 80 µg m^{-3} to < 18 µg m^{-3}). Most of this reduction is associated with the declining use of coal and oil (Viles 1996).

Over the last 10 to 20 years, very low levels of SO_2 have been coupled with a complex spatial pattern of NO_2 and PM_{10} pollution primarily from road traffic. Busy roads show high NO_2 and PM_{10} levels, but other (urban background) areas have very low concentrations (Oxford Airwatch).

Notwithstanding numerous restoration campaigns, stone deterioration in Oxford remains a very costly issue, despite these recent improvements in air quality. The decay path observed on these limestone walls typically comprises localized blistering and contour scaling of black crusts, together with an outer layer of the underlying stone, followed by stabilization of the exposed stone through the formation of a new black crust. This scaling forms a fragile dynamic equilibrium, which can be upset by cleaning or other processes which inhibit the 'healing over' of fresh surfaces.

LAB EXPERIMENTS ON LIMESTONE BLOCKS

Before using a Konica Minolta Vi9i™ object scanner to capture real-world stone surfaces, a laboratory-based accelerated weathering experiment was conducted to investigate the efficacy of the machine as a surface-monitoring tool, develop procedures on how best to use it, and test geostatistical methods/models for the analysis of surface change.

Two freshly cut ashlar blocks of Stoke Ground Base Bed limestone, measuring 18 cm × 24 cm × 10 cm deep were used in this study—labelled B1 and B2. Four brass survey pins were fixed to the side of the blocks as reference points. The blocks were then insulated with expanded polystyrene on five sides to leave one exposed 18 cm × 24 cm face through which salt, moisture, and temperature could be cycled. Stoke Ground Base Bed is a sorted oosparite limestone that has been regularly used as a building stone in Oxford for the original construction and continuing repair. It was chosen for this experiment to give a realistic comparison to the selected field sites in Oxford. Salt was chosen as the agent of decay in this study due to the largely physical and more aggressive deterioration associated with its internal crystallization.

The two blocks were placed face up in a commercial salt corrosion cabinet that allows the samples to be wetted with a fine mist of salt solution. The salts used in this experiment were a mixture of 5% by weight of sodium chloride and 5% by weight of magnesium sulphate. This combination of salts is representative of the mix commonly found in polluted urban environments (Smith et al. 2003).

The two blocks were wetted with a fine mist of the salt solution for 1 hour at the beginning of each 24-hour cycle of heating and cooling, with 10 hours at 20 °C followed by 10 hours at 40 °C, allowing 2 hours between each temperature change for heating and cooling. After 5, 20, 30 and 40 cycles, the stone samples were removed, allowed to cool at room temperature, and cleaned with a soft brush to remove any lightly held debris that in the real world would have been removed by wind or rain, and then 3D scanned. The scans were labelled according to their block number and suffixed with the number of weathering cycles (e.g., B2–20 is Block 2 after 20 weathering cycles).

The main consideration when using any measurement technology for detecting change at any scale is the use of reference points or control. Preferably these control points should be outside the area of interest and they should not move between successive measurements. Schaefer and Inkpen (2010) concluded that using shapes, in their case, table tennis balls, located in different planes relative to the scanned surface was found to be the most precise method in registering successive scans on small-scale weathered surfaces. However, this may be practical in controlled laboratory conditions but is not suitable for measurements in the field. Control for these blocks was twofold: the four brass pins, glued to the edge of each block and the surface of the polystyrene surrounding them. Polygon Editing Tools (PET), the vendor-specific software for operating the Konica Minolta Vi9i, was used to collect, manipulate, and register successive scans to the original unaltered surface.

In order to investigate change on the artificially weathered surfaces of the blocks, successive data sets have to be referenced to a common coordinate system. This was achieved by using the fresh unaltered surface for each block as a datum (i.e., 0 cycles). Despite attempts to ensure that the initial and subsequent scans on each block were parallel to the scanner, there were small errors in the setup. This was corrected using PET software. The initial scan for each block (B1–0, B2–0) was opened in PET and rotated around its x, y, and z axis until it was level and a point in the middle of these two surfaces was selected and its z value was set to zero. This initial scan will be referred to as the registration scan and the subsequent scans at 5, 20, 30, and 40 weathering cycles as comparative scans.

In PET, one can interactively select a minimum of three common points on two surface models and then let the software match them based on common surface geometry. This method was trialed by selecting the four brass pins as common points on two successive surface models, B1–0 and B1–5. Registration errors were acceptable, but as the software was fitting a surface to a surface and not using the selected points for reference, this technique has little or no value in studying changing surfaces. To work around this problem, the top surface of the surrounding expanded polystyrene was used as a constant unchanging reference datum.

To achieve this, the surface of the stone in the registration scan and the comparative scan was temporarily 'hidden' and not involved in the registration process. Three common points were selected on the remaining modelled polystyrene surfaces surrounding the stone and then the software did the matching based on these common surfaces. In PET, the registration scan does not move during this process. Rather, it is the comparative scan that is brought to it. After registration, the stone surface was switched back on. This method achieved very good registration results with an average error of 0.013 mm between scans for B1 and 0.012 mm for B2. In PET, the closer the error average is to zero, the more accurate the registration is. Each set of block scans was edited to remove the data on the surface of the surrounding polystyrene plus circa 10 mm around the edge of the stone surface. This was to reduce the 'edge effect' of the unconstrained artificially weathered sample block. This resulted in a surface of dimensions 213 mm x 158 mm for all Block 1 data and 215 mm x 157 mm for Block 2.

The registered data files were then exported from PET as ASCII files at a point density of 1 mm and imported into ESRI's ArcMap GIS software. A triangulated irregular network (TIN) was constructed for each data set using 3D Analyst (Figure 6.1)

A large number of indices have been applied to surfaces across a wide range of scales from millimetres to kilometres by geomorphologists. Most of these statistical constructs have concentrated on 2D linear transects, and the majority of small-scale weathering studies collect this data using a profile gauge (McCarroll & Nesje 1996). This involves pressing the gauge against a stone surface, removing it, and

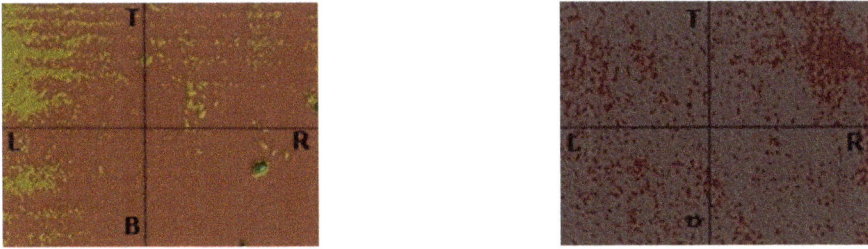

FIGURE 6.1   (Left) Location of profiles L-R and T-B on Block 1, (Right) Location of profiles L-R and T-B on Block 2.

then tracing the profile onto graph paper—this technique risks damaging the surface under investigation and collects a limited amount of data in two dimensions.

To investigate the value of this 2D linear approach on these 3D data sets, a 'digital' profile gauge was used on two transects L-R (left-right) and T-B (top-bottom) on all surfaces (see Figure 6.2). This was achieved using the 'Interpolate line' tool in ESRI's 3D Analyst extension. Despite the 2D nature of this data, it still provides some useful insights into how the surface of the stone is reacting to this artificial salt weathering. This 2D profilometry also shows the variation in the two test blocks' resistance to weathering. Although they are the same stone, weathered under the same conditions, Block 1 profiles show a continuing irregular surface expansion with little material loss up to 40 cycles, while Block 2 shows a switch from surface expansion to surface loss between 30–40 cycles. This is reflected in the mean heights calculated for all four profiles. When these values are plotted against the number of weathering cycles (Figure 6.2 (A))—both profiles on B1 show a continuous rise in the mean value after 5 cycles, while B2 shows a distinct fall on both profiles after 30 cycles. This data also shows that all profiles on both test blocks show a fall in the value of mean height from 0 to 5 cycles. This fall in mean height between 0 and 5 cycles is due to the initial infilling of small depressions with crystalline salt.

The weathering of stone surfaces, by chemical, physical, and biological processes, frequently results in an increase in surface roughness, and attempts have been made to use surface roughness as an indicator of the degree of surface weathering. Defined by McCarroll and Nesje (1996) as "often obvious but difficult to quantify satisfactorily", roughness means many things to many people.

The most widely used metric for surface roughness is root mean square (RMS) height. High RMS values indicate rough surfaces, as the surface heights over a given area show a large deviation from the mean value. RMS height is the standard deviation of heights above a mean for a given sample area (Shepard et al. 2001).

RMS height was calculated for each 3D surface (Figure 6.2 (B)). For B1 and B2, these results show that after 5 cycles both surfaces get smoother—for B1 RMS height falls from 0.119 mm to 0.116 mm and for B2 from 0.071 mm to 0.067 mm. For B1, the surface continues to get increasingly rougher at 20, 30, and 40 cycles through surface swelling. B2 shows a similar trend up to 30 weathering cycles; however, between 30–40 cycles, the RMS value increases only slightly. This is another indicator of the switch from surface expansion to surface loss on B2 between 30–40 cycles.

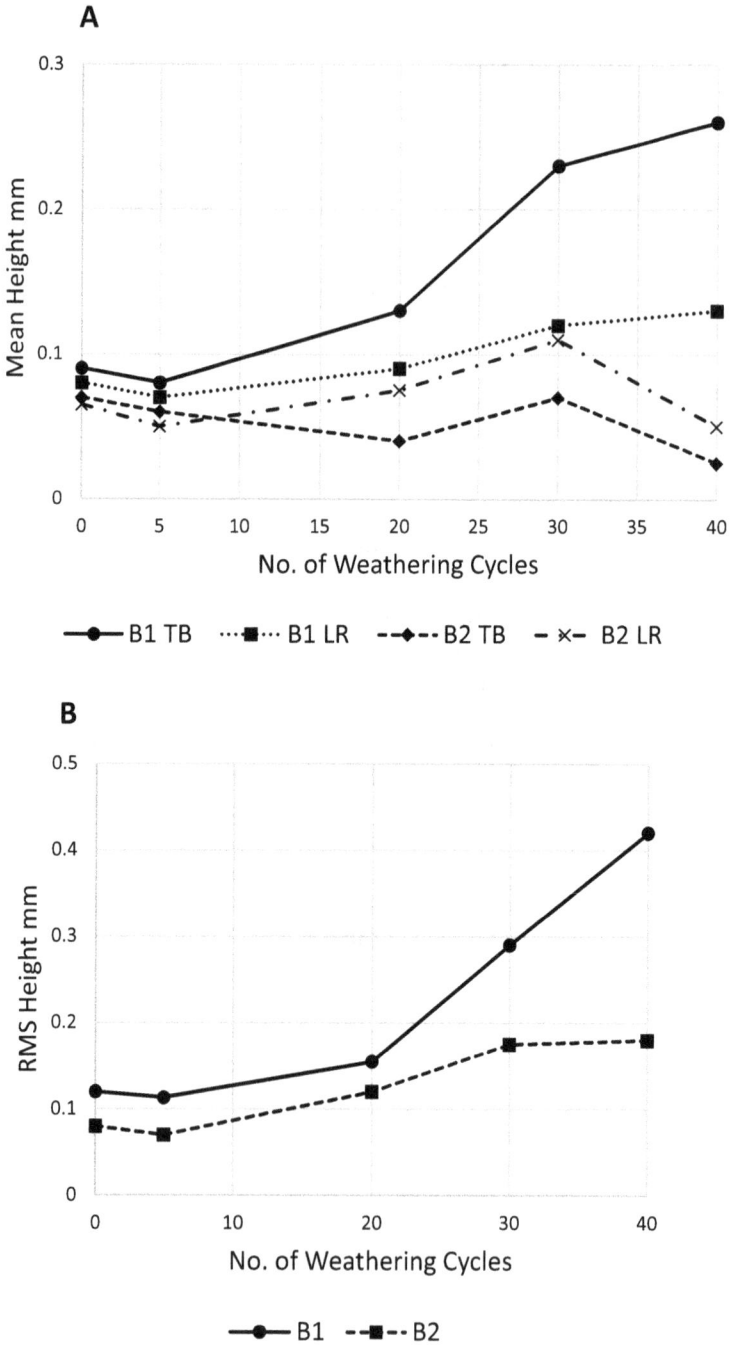

FIGURE 6.2 (A) Mean surface height (mm) for the four surface profiles B1 L-R, B1 T-B, B2 L-R, and B2 T-B plotted against the number of weathering cycles. (B) 2D RMS height mm of B1 and B2 plotted against the number of weathering cycles.

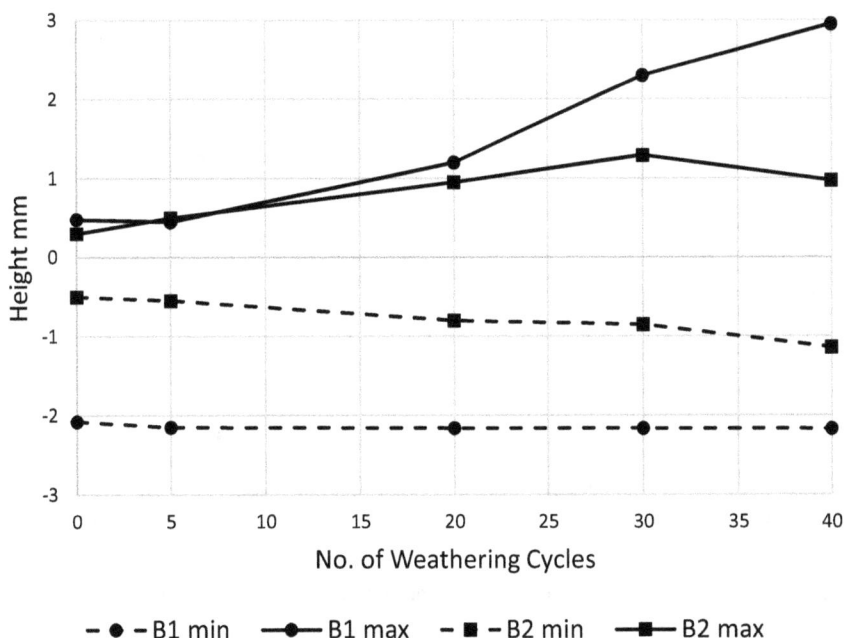

FIGURE 6.3 Plot of minimum and maximum height values (mm) for each Block DEM against the number of weathering cycles.

The conclusions drawn from the 2D digital profiles and 3D RMS height values on what is happening to the surface of the blocks during this artificial weathering process are further supported when the minimum/maximum height values from the DEMs are plotted for each block against the number of weathering cycles (Figure 6.3).

Block 1 shows very little change in minimum height throughout this whole process ranging from −2.08 mm at the start to only −2.17 mm after 40 cycles. However, the maximum height value rises steadily from 0.48 mm after 5 cycles to 2.95 mm at 40 weathering cycles—corroborating that the surface is predominantly swelling up to this point with minor material loss in the hollows. Block 2 shows a similar pattern up to 30 cycles, but from 30–40 cycles the minimum height on the surface increases from −0.85 mm to −1.14 mm and the maximum height falls from 1.29 mm to 0.97 mm—supporting the shift from a period where surface swelling is dominant with minor material loss in the hollows to a period where surface failure/detachment is the dominant process.

Using ESRI's ArcMap, an artificial digital reference plane (R.P.) was constructed and placed at the mean height of the original unaltered surface for both blocks. For Block 1 this was 0.072 mm and 0.042 mm for Block 2. The volume, in cm³, of the space below and above this reference plane with respect to the TIN surface was calculated for all surfaces using ESRI's 3D Analyst. Any increase in the volume below and decrease in the volume above this reference plane would correspond to the material being lost from the surface. Conversely, any decrease in the volume below

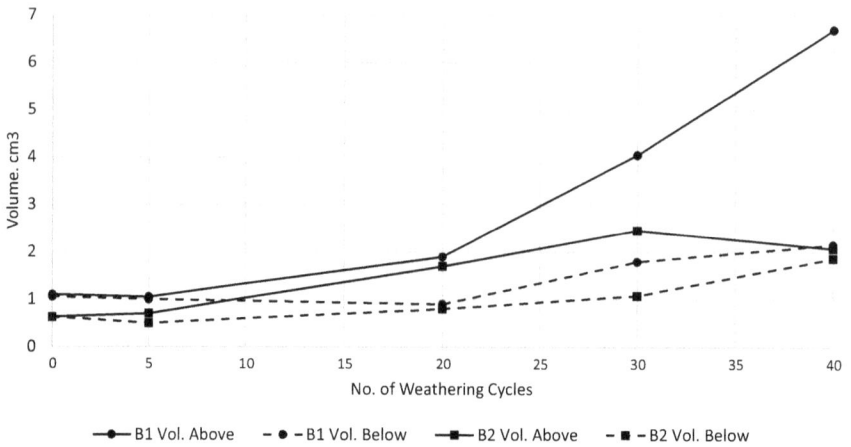

FIGURE 6.4   Block 1 and 2—Volume above/below reference plane versus number of weathering cycles, cm³.

and increase in the volume above this plane would indicate surface swelling or pore filling with salts.

The volume above/below the R.P. results for scans B1–0 and B2–0 validate the initial alignment of these registration scans (levelling) in PET with B1–0 having 48.8% below/51.2% above and B2–0 49.8% below and 50.2% above. The volume above/below the R.P. values was then plotted against the number of weathering cycles for each block (Figure 6.4).

Block 1 shows a similar trend to all the other data calculated on this surface—an initial fall in the volume below the R.P. between the start and 5 weathering cycles supports the premise that some pore filling is occurring during this early exposure to salt. However, it also reveals that a small amount, 0.07 cm³, of material was also lost from the surface above the reference plane at this stage. From 5 to 20 cycles there is little change in the volume below the R.P., and from 20 to 40 cycles this volume increases at a moderately linear rate to 2.16 cm³. This is a total loss of 1.1 cm³ of material below the R.P. The volume above the R.P. for Block 1 changes much more rapidly than the volume below. From its initial value of 1.11 cm³ at 0 cycles to 6.69 cm³ after 40 cycles, emphasizing that the dominant process occurring on this block after 40 cycles is surface expansion through salt crystallization.

The volume calculations with respect to the R.P. for Block 2 reveal a similar story to that of Block 1 up to 30 cycles. A small increase in the volume below the R.P. from 0 to 5 cycles followed by a moderately linear increase in volume from 0.63 cm³ at the start to 1.08 cm³ at 30 cycles. The volume above the R.P. for this same period shows a similar but not as large an increase to B1, with 0.63 cm³ at 0 cycles to 2.46 cm³ at 30 cycles. As with Block 1, the dominant process up to 30 cycles is surface expansion with some material being lost. However, things change between 30 and 40 cycles when the volume above the R.P. falls from 2.46 cm³ to 2.08 cm³, representing a loss of 0.38 cm³ of material and the volume below the R.P. increases from 1.081 cm³ at 30 cycles to 1.87 cm³ at 40 cycles—a loss of 0.79 cm³. This is a total loss of 1.17 cm³. Block 2 at this stage has clearly crossed a threshold where the surface has expanded to the point of failure and has now entered into a phase of material loss. It is interesting to note that the volume of material lost below the R.P. between cycles 30 and 40 is approximately double the volume lost above it.

VISUALIZING SURFACE CHANGE

Probably the most useful and understated aspect of capturing surface change in high-resolution 3D is the ability for researchers to see what is going on. This is especially true if the data is to be shared with non-scientific stakeholders. It is often difficult to explain mathematical constructs of what is happening to a surface. For example, showing a graph of increasing RMS height values calculated from a profile gauge to non-scientists would be time-consuming, but showing a series of color-coded contour DEMs to the same audience, they can easily see and get a feel for what is happening to the surface over time.

The graph in Figure 6.5 is a subsection of the T-B 2D digital profile for Block 1, and it shows the development of a blister on the surface approximately 40–60 mm along this profile.

However, this 2D view, informative as it is, does not tell the whole story. Figure 6.6 shows the same blister development but this time in 3D. These oblique views of the

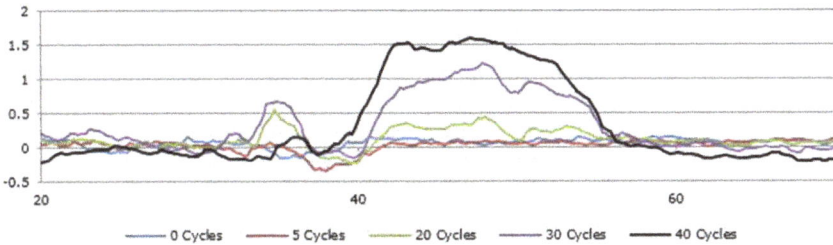

FIGURE 6.5   A subsection of 2D digital profile T-B Block 1.

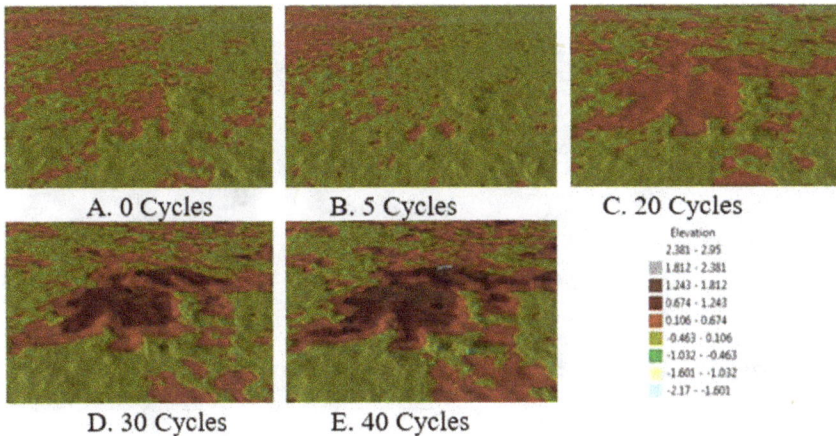

A. 0 Cycles   B. 5 Cycles   C. 20 Cycles

D. 30 Cycles   E. 40 Cycles

FIGURE 6.6   (A–E) 3D oblique views of a blister forming on the surface of Block 1. Scale in mm.

DEMs for Block 1 were generated in ESRI's Arc Scene, and they show a much more complex pattern of growth and the spatial extent of the blister in all directions.

This laboratory-based experiment confirmed the potential of this high-precision, noncontact, non-destructive technique as a very sensitive tool for measuring surface change. Its purpose was not to comprehend the salt weathering of limestone but to establish if significant and, more importantly, useful measures of change could be acquired using this method. With this in mind, it has been demonstrated that the technique can be used to capture changing surfaces at sub-millimeter detail, with a high degree of precision in 3D. This should give researchers in this field new insight into what is happening to a surface as weathering progresses. It also highlights that two stones, cut from the same block and weathered under the same conditions react in different timescales.

FIELD-BASED EXPERIMENTS—WORCESTER COLLEGE, OXFORD

A single ashlar block on the south-facing outer wall of Worcester College's Chapel was chosen for this study. Worcester College was founded in 1714, but there has been an institution of learning on the site since the late 13th century. While Worcester College is very close to the center of Oxford today, in the 18th century it was on the edge of the city. The Hall and Chapel, which flank the entrance to the College, were completed in approximately 1770. Overall the stonework on this wall, constructed from local Headington freestone (Sass & Viles 2010), appeared sound with surface patches of deterioration. The block chosen for the study had already lost approximately 50% of its original ashlar surface (Figure 6.7 (A)). Results from the lab-based study described show that individual blocks react differently to the same conditions, hence the patchwork appearance of this wall. All of these ashlar blocks will eventually decay; some are just further down that road than others.

The subject block was surveyed with a Konica Minolta Vi9i high-resolution object scanner in 2006, 2007, 2008, 2010, and 2011. This object scanner is primarily designed

FIGURE 6.7 (A) Photograph of the selected ashlar block, Worcester College. (B) The Konica Minolta Vi9i object scanner surveying the chosen block.

for use indoors (light levels <500 lux) and in normal daylight conditions the reflected natural light from the surface drowns out the returning laser measuring the surface. To improve this signal-to-noise ratio, the block under investigation was shaded with a dark umbrella (Fig 6.7 (B)). The same method outlined before for surface registration and DEM construction was used on these five data sets. A digital reference plane was placed on the surface of the first data set collected in 2006 and set to 0 mm, the interface between the red and yellow contours on the digital surface models in Figure 6.8.

The volume of the material above this reference plane and the volume of the void below it was calculated for all the data sets in cm^3. This data is presented in the graphs in Figure 6.9.

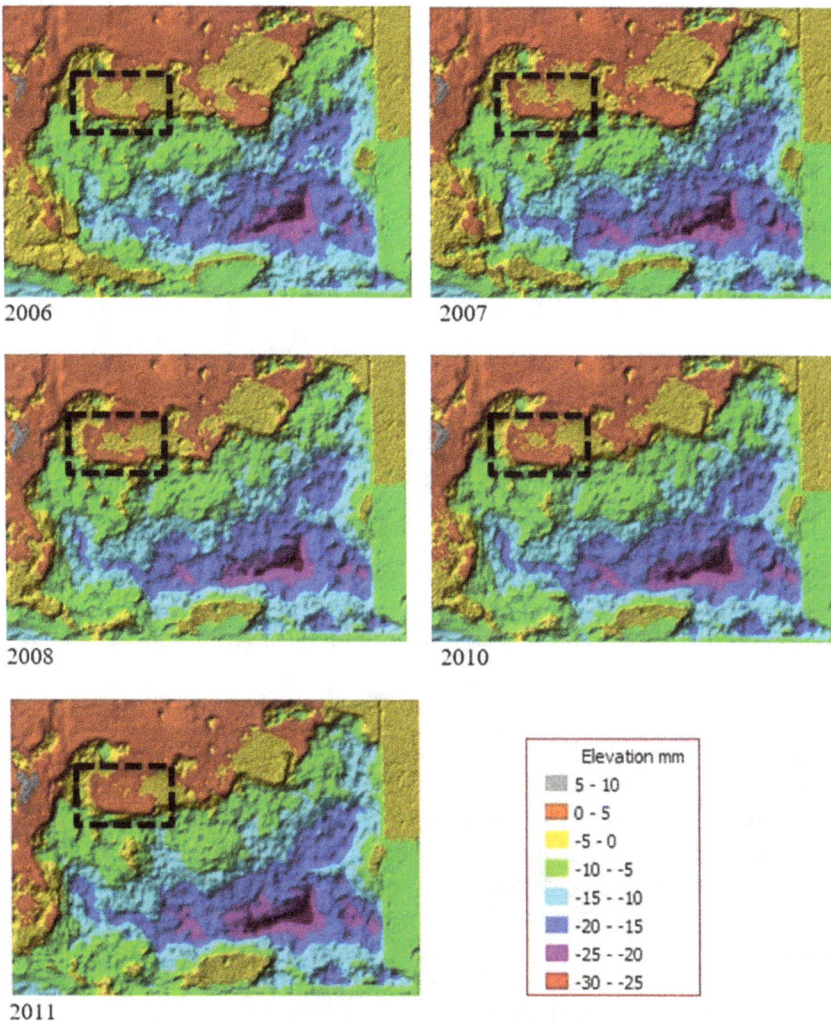

FIGURE 6.8   DEMs of the Worcester College block 2006–2011.

A

□ Vol. Above Ref. Plane cm3

B

□ Vol. Below Ref. Plane cm3

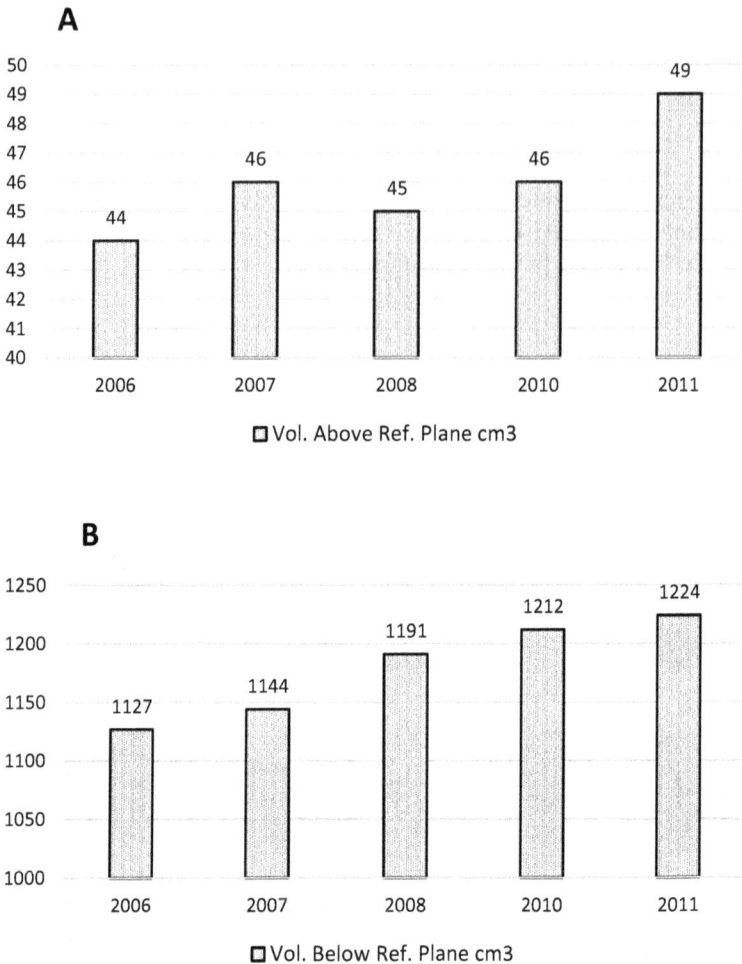

FIGURE 6.9   (A) Volume above R.P. and (B) volume below R.P on the Worcester Block, cm³.

The volume of the material above this plane increases from 2006 to 2007. It then falls in 2008 and increases again in 2010 with the biggest increase in 2011. The volume below the reference plane data shows an increase year on year with the biggest jump between 2007 and 2008—this corresponds to a decrease in material above it in the same period. This is mainly due to the loss of material highlighted in the solid black box in Figure 6.11—where an area of material, with its case-hardened surface still intact at the start of monitoring, swelled up from 2006 to 2007, becoming detached before the 2008 survey. Figure 6.10 is a series of 2D cross-sections generated in Arc GIS along the dash-dot line in Figure 6.11 and clearly shows the swelling of the above flake between 2006 to 2007 before its eventual loss in 2008. Although still attached, this process is repeated in the area highlighted in the dashed box on all the surface models in Figure 6.11.

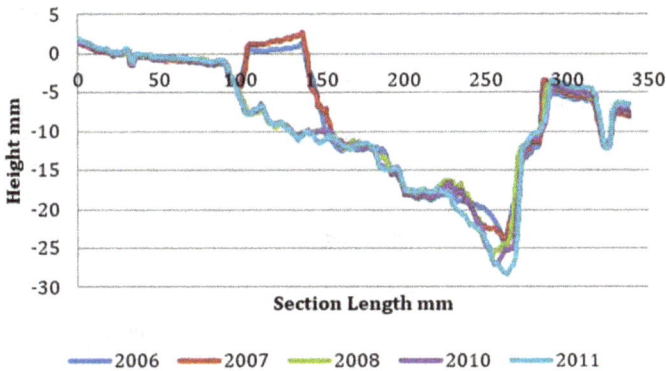

FIGURE 6.10   2D cross-sections along the dash-dot line in Figure 6.11.

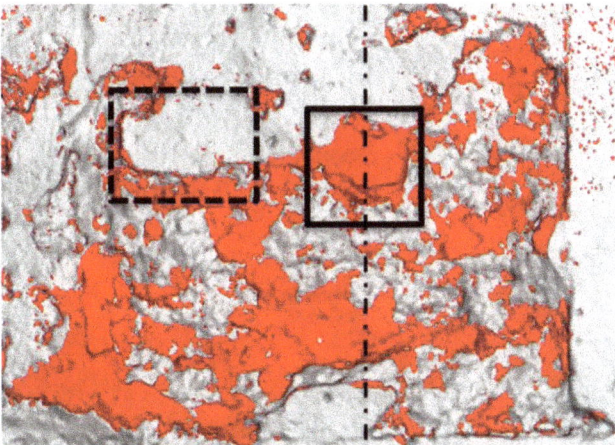

FIGURE 6.11   2006–2011 surface difference on the selected stone at Worcester College. All the material highlighted in red was lost during this time period.

This high-resolution monitoring and detailed analysis of the 3D models constructed from the data collected at the Worcester College study site have given valuable insight into this mechanism and spatial distribution of decay. This data reveal that once the hardened surface crust is removed or breached, the majority of decay proceeds along the path of granular disintegration within the main hollows, while the hardened surface crust fails by sub-surface swelling, predominantly along its exposed edge. This is underlined in Figure 6.11, which shows that the vast bulk of material lost over the monitoring period on this stone, highlighted in red, is concentrated in the main cavern and its exposed edges and the minor hollows surrounding it.

FIELD BASED EXPERIMENTS—NEW COLLEGE WALL, OXFORD

An example of this dynamic equilibrium is found on the 14th century, or maybe earlier, Headington freestone walls forming the exterior of the cloisters at New College

Lane. The middle and upper parts of the wall, which are covered by black crust, still show the original masons' tool marks and remain generally intact (Figure 6.12).

In contrast, at the base of the wall, there is evidence of former shallow deterioration and recession in the form of scaled and re-crusted surfaces but also large areas of original and secondary crust that are now subject to widespread undermining by actively eroding cavernous hollows characteristic of the physical disaggregation caused by salt weathering. This image also shows areas where the guttering has failed, letting rainwater run down the surface of the wall and inhibiting crust formation.

To measure the rate and spatial distribution and to investigate the causes of this cavernous weathering, 3D scanning of an individual hollow (black square, Figure 6.14) and the surrounding black crust were surveyed using the Konica Minolta Vi9i object scanner in 2007, 2008, and 2010. (Figure 6.13).

FIGURE 6.12 Photograph of part of the New College wall.

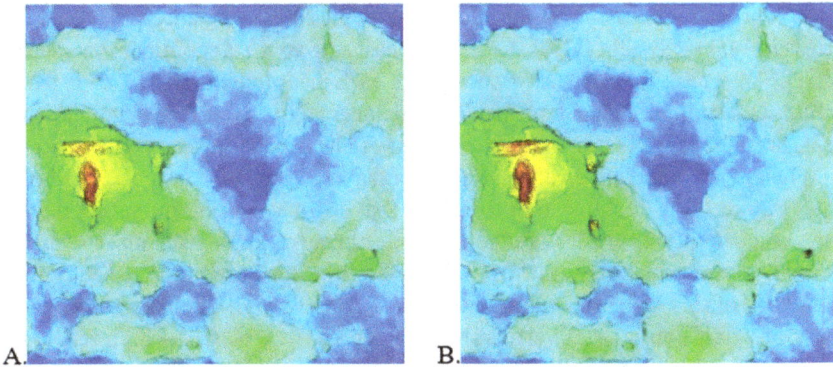

FIGURE 6.13 (A) DTM of stone hollow in 2007, (B) in 2010. Blue=high, Red=low.

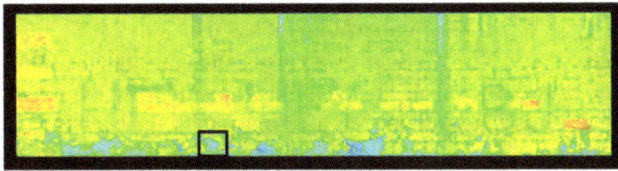

FIGURE 6.14 DEM of the 15 m section of wall in New College Lane. This is color-coded with intensity values (I) collected by the laser scanner—that is, how much of the laser light sent out by the survey equipment to measure the surface returns to the machine. Red/orange = low I—Green= medium I and Blue= high I.

Over this monitoring period on New College Lane, the cavernous hollow lost 438.9 cm^3 of material in a spatially patchy pattern associated with the differential deepening of a number of 'hot spots' (Figure 6.13). Deepening of the hollow was also accompanied by lateral enlargement as parts of the surrounding black crust peeled away, resulting in a net increase in the horizontal surface area of the hollow of 6.44% over the entire period. Assuming that the mass loss is consistent over the entire hollow, the mean recession rates in the hollow (calculated with respect to an arbitrary reference plane fitted over the oldest, most stable crusted surface) range from 1.49 mm pa (2008–2010)—2.05 mm pa (2007–2008), with a mean over the entire period 2007–2010 of 1.73 mm pa. In this case, much of the loss was explained by the creation of a number of new, small hollows towards the base of the section and the lateral growth of existing hollows resulting in a concomitant loss of crusted material above the artificial reference plane. Whilst the recession rate calculations give highly conservative values, they are still around 25–70 times those observed in the 1980s and 1990s on the Portland limestone of St. Paul's Cathedral (Inkpen et al. 2012) and on many other limestone buildings in Europe (Sabbioni 2003). The persistent and significant mass loss and surface recession in the hollow also contrast markedly with the general lack of change observed on the surrounding black-encrusted surface, apart from small surface blisters and limited exfoliation around the edges of the hollow.

A 15 m section of this wall was also surveyed in 2006 using a Leica HDS™ 3000 terrestrial LiDAR system (TLS). As well as measuring an XYZ coordinate on a surface, this TLS also returns an intensity value (I) for each measured point—which is the percentage of laser light that is returned to the scanner from the surface being measured. As its name implies, it is a measurement of how reflective the surface is at that point.

When these reflectance values are mapped and colour-coded onto a DEM of the wall (Figure 6.14), it shows that the areas that still have a dense coating of black crust and, therefore, are less reflective (red/orange/yellow) cover most of its surface. The cavernous hollows of freshly eroded limestone are predominantly found at the base of this wall, which have no black crust have high-intensity values (blue). There are also a number of vertical linear features, which exhibit high reflectance values at the top of the wall and fade into a slightly less reflective surface (green) down the wall. These are areas where the guttering along the top of the wall has broken, letting rainwater wash down the surface, reducing crust growth while at the same time eroding the surface through dissolution.

2D resistivity profiles were collected by Sass and Viles in 2006 and 2007 on both the Worcester and New College monitoring sites discussed previously. Using a Geolog 2000 with specialised shielded cabling, moisture distributions were mapped up to 40 cm inside the stone along a 2 m high transect of the New College Lane wall, from ground level, passing through an adjacent cavernous hollow. Surface moisture measurements were taken monthly from Sept. 2007 to Dec. 2008 using a Protimeter on three black-crusted areas and one cavernous hollow. Released debris was also collected on a monthly basis below nearby cavernous hollows between August 2007 and March 2008. Debris collected during this period confirms that material is being eroded from the hollow (c. 0.1 to 2 cm^3 per month), although the amount collected vastly underestimates the amount produced, because of losses to wind and runoff (Sass & Viles 2010)

This data set fits a model whereby, under previously polluted conditions of high atmospheric sulphur and particulate loadings, limestone surfaces were rapidly blackened and encrusted with gypsum, formed mainly as a reaction product between $CaCO_3$ and atmospheric SO_2 on moist stonework. These crusts formed more rapidly than they could be removed by limited surface water washing, but periodically areas of crust could be lost through blistering and surface scaling. This most probably resulted from the sub-surface crystallization of gypsum and other salts washed into the stonework and concentrated below the surface by periodic wetting and drying. Within the context of New College wall, blistering and surface loss are seen to concentrate in a zone towards the top of the capillary fringe, around 0.5–1 m above pavement level, where salts washed in from the surface are supplemented by others brought upwards in rising groundwater—in a pattern that conforms with recent theoretical models of salt weathering (Hall et al. 2010). Historically, however, it would appear that surface scaling was typically followed by the rapid re-formation of a new black crust that effectively 'healed' the scars left by contour scales before they could grow into cavernous hollows. This pattern of periodic surface loss, followed by rapid stabilization is, as indicated, reflected in the serial black crusts observed on many ancient Oxford walls but can also be actively observed on similar limestone in cities such as Budapest. Such cities act as present-day analogs for the previous pollution conditions of Oxford and also highlight the importance of high levels of surface particulate deposition (dust) in catalysing the rapid formation of gypsum (Smith et al. 2003).

In the present-day absence of high levels of atmospheric sulphur dioxide and rapid surface dust deposition in Oxford, it would appear that once areas of black crust are breached, they are now unlikely to heal over but, instead, may become the focus for rapid cavernous weathering. The rapidity of this weathering can possibly be linked to changes in the pattern of sub-surface moisture conditioned, in turn, by the growth of the cavernous hollow itself. This is demonstrated by the 2D resistivity surveys and surface Protimeter measurements (Sass & Viles 2010). The 2D resistivity profiles reveal generally dry conditions below the hollow, damp conditions within it and a drier near-surface zone and extremely wet deep sub-surface conditions beneath the upper, black, encrusted stonework above the hollow. Monthly Protimeter measurements confirm dry surface conditions throughout the year under the hollow; variable wet/dry conditions within the hollow, and seasonally wet/dry conditions on the black encrusted stonework above the hollow. Repeated moistening and drying within the hollow is particularly conducive to near-surface weathering, through alternating dissolution—crystallization and/or hydration—dehydration of embedded salts, and is reflected in the characteristic flaking and granular disaggregation observed within the hollows. In turn, prolonged wetting beneath the surface seal of the original black crust would have aided the accumulation over many centuries of a reservoir of deeply penetrated salt that can continue to drive decay despite the ongoing loss of salt-rich debris from within the hollows.

CONCLUSION

Replacing 2D cross-sections, traditionally collected with a profile gauge to monitor stone decay, with high-resolution 3D scanning technologies has greatly increased the amount of data collected on a weathered surface and allowed researchers to visualise them through the construction of digital elevation models (DEM), refining the understanding of rapid weathering in limestone in the built environment and its spatio-temporal distribution.

The underlying conditions that favour rapid decay must, however, have been in place throughout much of the lifetime of the stones above but were controlled in the past by the self-limiting process of rapid crust formation. It appears that it is the change in the pollution regime that has shifted the system across a threshold from a state of dynamic equilibrium to one of positive feedback. Albeit, one that is exploiting a long history of internal salt accumulation and sub-surface weakening. This has resulted in the onset of a potentially catastrophic loss of structural integrity in the New College wall that has stood for over 600 years and, albeit at a smaller scale, the surface deterioration on the walls of Worcester College. These findings, therefore, suggest a more complex relationship between air pollution and limestone deterioration on historic walls than has been proposed in previous dose/response studies. Improving air quality may, under certain circumstances, threaten rather than benefit cultural heritage constructed in limestone as the removal of semi-protective crusts produced by air pollution may open the stone to attack by other agents of deterioration.

REFERENCES

Brimblecombe, P. and Grossi, C.M. 2008. Millennium-long recession of limestone facades in London. *Environmental Geology* 56, 463–471.

Brimblecombe, P. and Grossi, C.M. 2009. Millennium-long damage to building materials in London. *Science of the Total Environment* 407, 1354–1361.

Cardell, C., Benavente, D. and Rodriguez-Cordillo, J. 2008. Weathering of limestone building material by mixed sulfate solutions. Characterization of stone microstructure, reaction products and decay forms. *Material Characterization* 59, 1371–1385.

Diaz-Andreu, M., Brooke, C., Rainsbury, M. and Rosser, N. 2006. The spiral that vanished: The application of non-contact recording techniques to an elusive rock art motif at Castle rigg stone circle in Cumbria. *Journal of Archaeological Science* 33, 1580–1587.

Eggleston, S., Hackman, M.P., Heyes, C.A., Irwin, J.G., Timmis, R.J. and Williams, M.L. 1992. Trends in urban air pollution in the United Kingdom in Recent Decades. *Atmospheric Environment* 26, 227–239.

Fenger, J. 1999. Urban air quality. *Atmospheric Environment* 33, 4877–4900.

Hall, C., Hamilton, A., Hoff, W.D., Viles, H.A. and Eklund, J. 2010. Moisture dynamics in walls: Response to microenvironment and climate change. *Proceedings of the Royal Society A* 467, 194–211.

Inkpen, R. 2007. Interpretation of erosion rates on rock surfaces. *Area* 39.1, 31–42.

Inkpen, R., Viles, H. and Moses, C. 2012. Modelling the impact of changing atmospheric pollution levels on limestone erosion rates in central London, 1980–2010. *Atmospheric Environment* 61, 476–481.

McCarroll, D. and Nesje, A. 1996. Rock surface roughness as an indictor of degree of rock surface weathering. *Earth Surface Processes and Landforms* 21.

Pope, G., Meierding, T. and Paradise, T. 2002. Geomorphology's role in the study of weathering of cultural stone. *Geomorphology* 47, 211–225.

Rodriguez-Navarro, C. and Sebastian, E. 1996. Role of particulate matter from vehicle exhaust on porous building stones (limestone) sulphation. *The Science of the Total Environment* 187, 79–91.

Sabbioni, C. 2003 Mechanisms of air pollution damage to stone. In Brimblecombe, P. (ed.) *The effects of air pollution on the built environment*, Imperial College Press: London, pp. 63–106.

Sass, O. and Viles, H. 2010. Wetting and drying of masonry walls: 2D-resistivity monitoring of driving rain experiments on historic stonework in Oxford, UK. *Journal of Applied Physics* 70, 72–83.

Schaefer, M. and Inkpen, R. 2010. Towards a protocol for laser scanning of rock surfaces. *Earth Surface Processes and Landforms* 35, 147–423.

Searle, D.E. and Mitchell, D.J. 2006. The effect of coal and diesel particulates on weathering loss of Portland Limestone in an urban environment. *Science of the Total Environment* 370, 207–223.

Shepard, M., Campbell, B., Bulmer, M., Farr, T., Gaddis, L. and Plaut, J. 2001. The roughness of natural terrain: A planetary and remote sensing perspective. *Journal of Geophysical Research* 106, 32777–32795.

Smith, B.J., Srinivasan, S., Gomez-Heras, M., Basheer, P.A.M. and Viles, H. 2011. Near-surface temperature cycling of stone and its implications for scales of surface of deterioration. *Geomorphology* 130, 76–82.

Smith, B.J., Török, A., McAlister, J.J. and Megary, Y. 2003 Observations on the factors influencing stability of building stones following contour scaling: A case study of oolitic limestones from Budapest, Hungary. *Building and Environment* 38, 1173–1183.

Turkington, A.V., Martin, E., Viles, H. and Smith, B.J. 2003. Surface change and decay of sandstone samples exposed to a polluted urban atmosphere over a six-year period: Belfast, Northern Ireland. *Building and the Environment* 38, 1205–1216.

Viles, H.A. 1996. 'Unswept stone, besmeer'd by sluttish time': Air pollution and building stone decay in Oxford, 1790–1960. *Environment and History*, 359–372. White Horse Press.

Viles, H.A. 2001. Scale issues in weathering studies. *Geomorphology* 41, 63–72.

7 Unpiloted Airborne Laser Scanning of a Mixed Forest
A Case Study from the Alps, Austria

*Michal Gallay, Ján Kaňuk, Carlo Zgraggen,
Benedikt Imbach, Ján Šašak, Jozef
Šupinský, and Markus Hollaus*

INTRODUCTION

Unpiloted airborne laser scanning (ULS) mounted on a multirotor or a helicopter platform combined with low-altitude flight and relatively slow speeds produce point densities that are orders of magnitude greater than traditional airborne laser scanning. The ULS of forests has also produced point clouds with densities equivalent to terrestrial LiDAR scanning (TLS), which has its drawbacks in the occlusion of tree digitization and the limited efficiency and mobility of the TLS system setup on the ground (Wang et al., 2019). ULS laser scanning coupled with wide-scan angles produces point densities that can resolve individual tree and branch structures similar to those collected by TLS (Morsdorf et al., 2017; Wieser et al., 2017; Kellner et al., 2019; Brede et al., 2019). Low-altitude flight also reduces the impact of Global Navigation Satellite Systems (GNSS) positioning uncertainties that can increase with the distance between the sensor and the terrain. Recent developments have shown that the application of ULS is possible even under the forest canopy providing accurate and ultra-detailed point clouds (Hyyppä et al., 2020). Because drone flight is predominantly under the control of the investigator and is normally one order of magnitude less costly than traditional airborne laser scanning, flight plans can be developed to collect high-density measurements in novel ways that enable hypothesis testing or to evaluate the impact of various data collection strategies on remote sensing measurements.

A summary of current lightweight laser scanners suitable for UAS and their key properties is listed in Table 7.1. The majority of these systems, including all of the RIEGL units, are mechanical—in that they rely predominantly on rotating mirrors to emit and collect the returning laser. The rest of the systems are lighter and smaller,

DOI: 10.1201/9780429327575-7

TABLE 7.1

Summary of Lightweight Laser Scanners Suitable for UAS-Based on Manufacturer Specifications

Manufacturer/Model	Weight (kg)	Range Accuracy (cm)	Beam Divergence (mrad)	Laser Wavelength (nm)	Max. Measurement Rate (kHz)	No. Returns	Max. Measurement Range (m)	Horizontal FOV (deg.)
RIEGL/VUX-120	2	1	0.4	1550	1800	Multiple	720	100
RIEGL/VUX-240	3.8	2.0	0.35×0.35	1550	1500	Multiple	350	75
RIEGL/VUX-1	3.5	1.0	0.5×0.5	1550	500	Multiple	170	330
RIEGL/VUX-1HA	5.0	0.5	0.5×0.5	1550	1000	Multiple	120	360
RIEGL/VUX-1LR	3.5	1.5	0.5×0.5	1550	750	Multiple	215	330
RIEGL/miniVUX-1	1.55	1.5	1.6×0.5	905	100	5	150	360
RIEGL/miniVUX-3	1.55	1.5	1.6×0.5	905	200	5	330	360
Velodyne/Puck LITE	0.59	3	-	905	600	2	100	360
Velodyne/HDL-32E	1.0	2.0	3.0×1.5	905	695	2	100	360
Velodyne/Velarray h800	1.0	3	-	905	400	1	200	120
Ibeo/LUX 4L	1.0	10	1.4×1.4	905	18.5	3	200	110
Hokuyo/UTM-30LX	0.37	3.0–5.0	-	905	4.3	1	270	270
Sick/LD-MRS 420201	0.77	10.1	1.4 × 1.4 × 2.8	905	-	3	300	110
Hesai/Pandar40	1.46	2	3	905	720	2	200	360
Hesai/Pandar64	1.52	2	3	905	720	2	200	360
Ouster/OS2-32	1.1	3–10	1.6	865	655	1	240	360
Ouster/OS2-128	1.1	3–10	1.6	865	2500	1	240	360
Quanergy/M8-Ultra	0.9	3	-	905	420	3	200	360
Livox/Mid-40	0.7	2	-	905	100	1	260	98

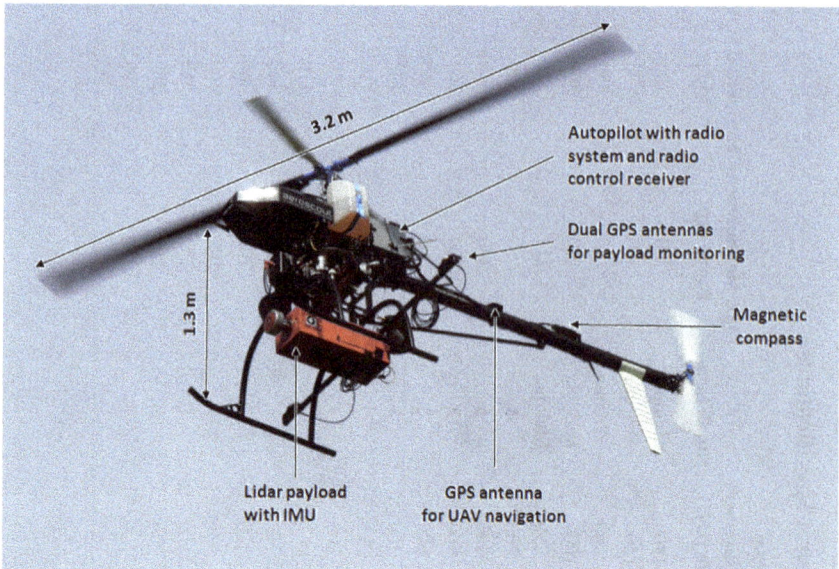

FIGURE 7.1 The unpiloted helicopter system Scout B1–100 by Aeroscout GmbH with a laser scanning payload at the site near Düns, Vorarlberg, Austria.

and their primary application is in advanced driver assistance systems (ADAS) and autonomous navigation (Lambert et al., 2020). They use a rotating array of laser emitters comprising 16 to 128 laser light emitters (channels) except the Livox and Velodyne Velarray devices. The two Ouster models are unique with their mechanically rotating, multi-beam flash LiDAR, which produces structured point clouds. The most recent advances in ULS use a solid-state LiDAR with no rotating components, such as in the Livox MID-40 or Velodyne Velarray H800, and in the near future, these solid-state scanners are likely to outperform those based on mechanical rotation. The performance of five of these ULS systems listed in Table 7.1, over three different kinds of landscapes are compared in Hu et al. (2021).

This case study from the Central Eastern Alps of Vorarlberg, Austria, used an unpiloted helicopter system, Scout B1–100, that was equipped with a VUX-1 LiDAR system and demonstrates the use of the data acquired in assessing local geomorphology, tree segmentation, and 3D modelling of solar irradiation with a ULS (Kaňuk et al., 2018) (Figure 7.1). The LiDAR system used in this research, although relatively old, was chosen because of its measurement accuracy, the number of laser returns and on-the-fly full-waveform processing. Additional results from this project can be found in Bruggisser et al. (2019).

AREA OF INTEREST

In this study, the LiDAR data acquisition covered an area between the villages of Düns and Dünserberg, Vorarlberg, Austria, with central geographic coordinates 47°13'20.83" N and 9°43'42.28" E (WGS84) (Figure 7.2). The total area mapped

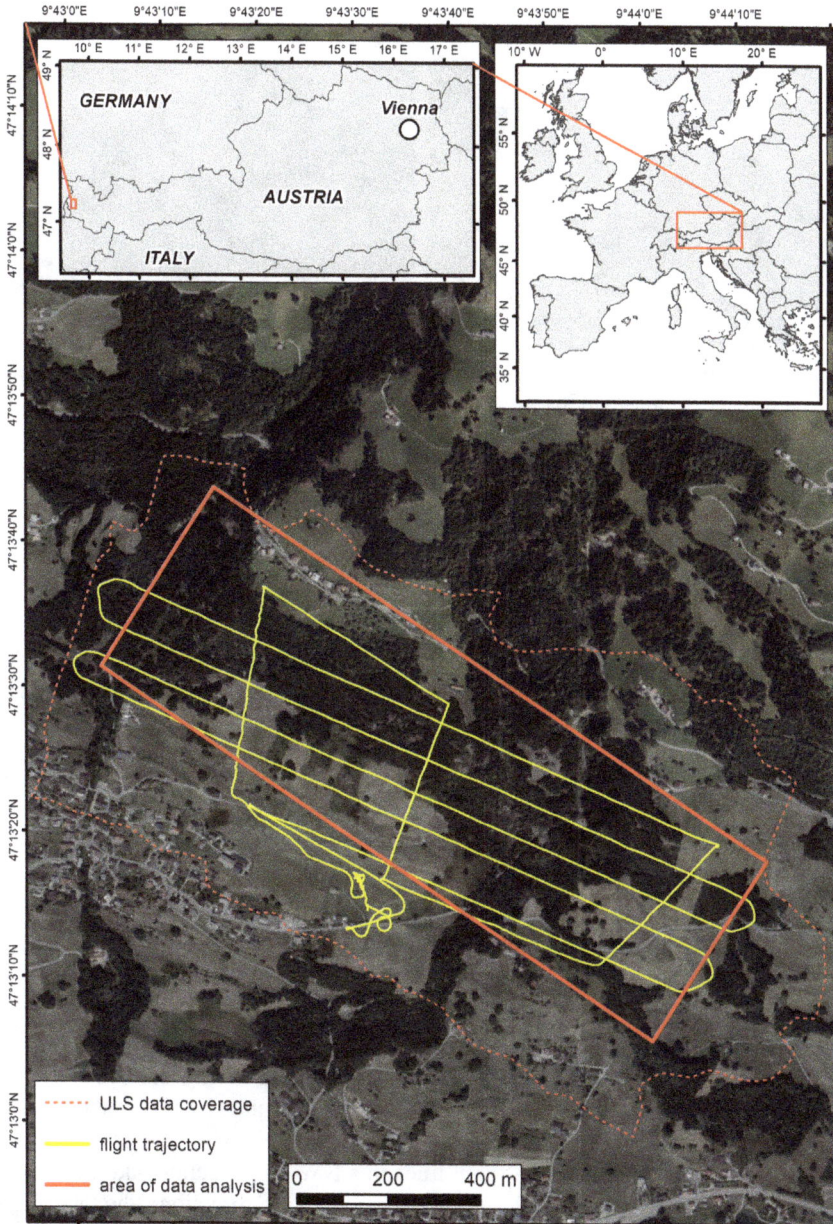

FIGURE 7.2 Location of the area of interest with flight lines, data coverage, and area of interest used in the analysis.

was 1.25 km², and this study focuses on the area outlined by the large red rectangle and covers 0.675 km² (1500 x 450 m). The area of interest is predominantly covered by mixed forest, dominated by coniferous trees (mainly spruce), and the elevation ranges from 800 to 1140 m.a.s.l. The trees grow on steep slopes (20–50°) where

FIGURE 7.3 Views of the laser-scanned steep slope near Dünserberg provide impression of a challenging environment for the ULS mission.

shallow landslides also occur (Figure 7.3). The mean canopy heights are 11.5 m. These topographic parameters made it challenging to conduct a ULS survey and required careful mission planning. Flight permission was approved by the Austrian aviation authority, Austro Control, and the survey was carried out on the 30 May 2017 just before midday. The weather was sunny, with a light breeze of 4 m.s^{-1} and air temperature of 25 °C. With such a heavy UAS, site accessibility and a suitable takeoff/landing location are important aspects of a successful survey mission. The area of interest was easily accessible by a local asphalt road so that all equipment needed for the flight could be transported by a van adjacent to the place of takeoff in the central part on a meadow.

ULS FLIGHT MISSION AND DATA PROCESSING

Aeroscout GmbH manufactured the helicopter and integrated it with the LiDAR system. The details of the technology used are summarized in Gallay et al. (2016) and Kaňuk et al. (2018). The laser scanner is a time-of-flight, pulse-based system, emitting infrared pulses of 1550 nm wavelength with a maximal pulse repetition frequency of 500 kHz.

The location, attitude, and orientation of the scanner during data collection is recorded by an Oxford xNAV550 inertial measurement unit (IMU) combined with two GPS antennas. The data was collected in a single flight with one set of orthogonal flight lines, comprising seven flight lines in a NW–SE direction and three flight lines, approximately at right angles to these, in a NE–SW direction. The total flight time for the survey was 48 minutes with a nominal flight altitude of 110 m above ground level and average flight speed of 6 m·s^{-1}. Given the scanners beam divergence of 0.5 mrad, the scan angle was limited to 90° and the scan density ranged from 4 cm between measurements at the canopy of the tallest trees to 6 cm open ground.

During the autonomous part of the flight, the flight control system maintained stable control of the aircraft and sensors. For example, during a representative flight line, the standard deviation of flight speed was 0.06 m.s$^{-1,}$ and the accuracy in the pitch, roll, and heading axes was 0.05°, 0.06°, and 0.09°, respectively. The quality of the post-processed flight trajectory defines the absolute accuracy of the ULS point cloud. This post-processing was performed in NAVsolve software by OXTS. The IMU and GPS data from the onboard ULS payload were integrated with the GPS base station recordings, resulting in an absolute accuracy of flight trajectory of 2–8 mm in WGS84 coordinates.

SUPPLEMENTARY DATA FROM UAV CLOSE-RANGE PHOTOGRAMMETRY AND TERRESTRIAL LASER SCANNING

In addition to mapping by ULS, two other methods of high-resolution mapping were used to compare this data—drone photogrammetry and TLS.

Close-range drone-based photogrammetry was used to derive a point cloud and a colour orthoimage of the study site. Overlapping images, collected using a DJI Phantom 4 UAV, were processed in Agisoft Metashape software, producing a point cloud and an orthoimage. The average point density over this UAV mapped 0.56 km^2 area was 187 points.m^{-2}. The average ground sampling distance (GSD–average pixel size) of the orthomosaic was 4.5 cm. For comparison with ULS and UAV point clouds, a small area was also surveyed using a RIEGL VZ-1000 terrestrial LiDAR system (TLS) system. This long-range, tripod-mounted scanner was used to collect detail on the slope and trees from three locations. These three data sets were then registered (joined together) using RiScanPro software with a combined error of 8 mm. The UAV and TLS point clouds were then co-registered with respect to the ULS data using RiScanPro software and multi-station adjustment with an accuracy of 80 mm and 95 mm, respectively. No GPS ground control points were used.

CHARACTERISTICS OF THE ULS POINT CLOUD

The ULS point cloud contained 432 million unique laser returns with an average density of 345.74 points·m^{-2}, ranging from 100 up to 2000 points·m^{-2} (Figure 7.4). In comparison, the average point density of traditional, piloted, airborne LiDAR data collected by the State of Vorarlberg is circa 30 points·m^{-2} (Bruggisser et al., 2019). The mean density of first returns over the forested section of the study area was 1850 points·m^{-2} with 105 ground points·m^{-2} on average. Laser pulses

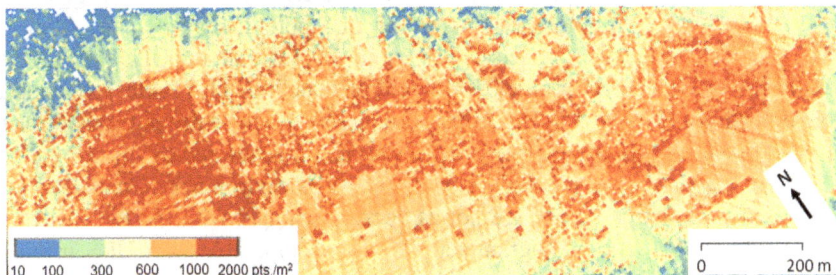

FIGURE 7.4   Point density of the ULS point cloud calculated in 5 x 5 meter cells.

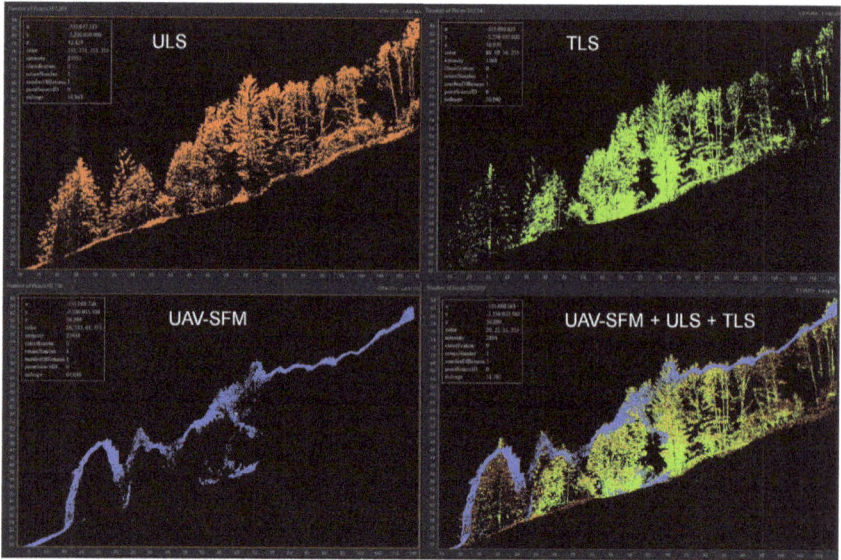

FIGURE 7.5 Vertical cross-section of a forest mapped by unpiloted airborne LiDAR (ULS), unpiloted close-range photogrammetry (UAV-SFM), and terrestrial LiDAR (TLS). The profile line is located in Figure 7.7.

emitted from the ULS reflect from objects both on and above the ground surface (e.g., vegetation, buildings, and bridges). One emitted laser pulse can return to the LiDAR sensor as one or a number of returns. Any emitted laser pulse that encounters multiple reflection surfaces as it travels in the direction of the ground is split into as many returns as there are reflective surfaces. The first returned laser pulse is the most important return and will come from the highest feature in an area like a treetop or the top of a building. However, the first return can also represent the ground, in which case only one return will be detected by the ULS. Multiple returns can be used to detect the elevation of several objects within the laser footprint of an outgoing laser pulse. These intermediate returns are generally used to determine vegetation structure, with the last return used for bare-earth (vegetation free) terrain models.

The level of detail captured by ULS and the difference in scanning geometry with respect to the UAV photogrammetry and TLS point clouds are clear from the individual and combined cross-sections in Figure 7.5. Overall, the ULS successfully mapped the whole vertical profile of the site. The benefit of TLS is clear in the near ground portions of the forest, but the upper parts of the trees are not satisfactorily captured due to occlusion by other tree trunks and branches and this limits the use of TLS for mapping wide areas of a complex forested environment. Conversely, the bottom sections of tree trunks are not as densely scanned with the ULS as compared to the TLS. UAV photogrammetry is only comparable with ULS in open, non-vegetated areas and only captured the vegetation canopy with very limited data on tree structure or terrain under the forest canopy.

Vegetation cover was categorized by the use of filtered and classified returns of the ULS data. Figure 7.6 shows a cross-section and perspective view of the point cloud coloured by vegetation, ground, and buildings and illustrates the high level of detail acquired using this system. The height difference between the first and last ground return (normalized height) presented in Figure 7.7 shows the distribution of vegetation canopy height. Areas with the darkest blue indicate trees taller than 30 m.

Being able to map the terrain below any vegetation is not only vital for calculating vegetation height, but it can also be used to identify geomorphological features and clues to the processes that formed them. The ULS ground returns were used to derive a digital terrain model (DTM) of the study area. The series of maps in Figure 7.8, produced using GRASS GIS software, show some basic geomorphometric parameters derived from the DTM. The central part of the area comprises undulating terrain under the forest (Figure 7.7). This is the result of shallow landslides. Inclination of some the trees in the cross-section of Figure 7.6 also indicate historic soil movement. This DTM can also be used to model environmental phenomena directly affected by the topography such as snow thaw, solar irradiation, and species distribution.

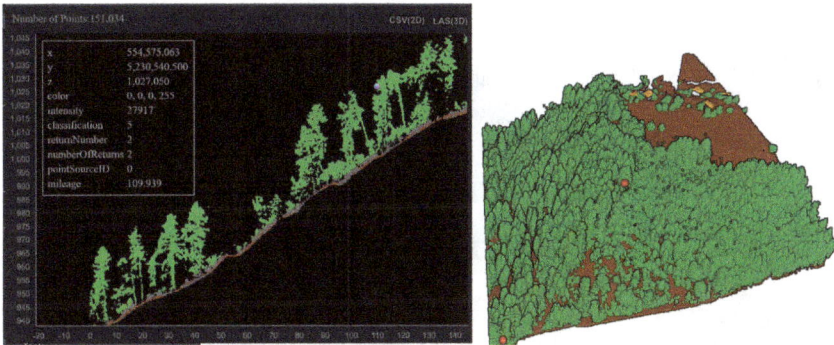

FIGURE 7.6 Cross-section and 3D perspective view of the classified UAS LiDAR point cloud. The location of cross-scetion is marked with red dots in perspective view, and this transect is also shown in Figure 7.7 (red line).

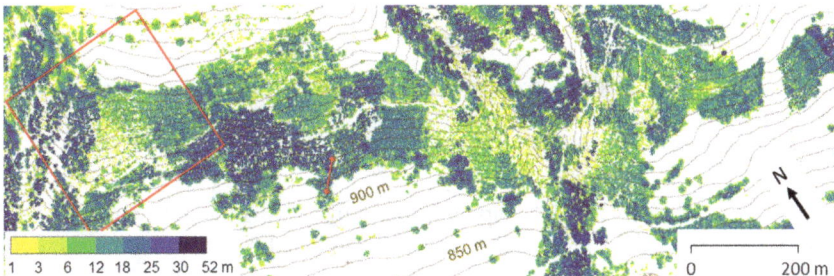

FIGURE 7.7 Vegetation canopy height model derived from points higher than 1 meter above ground level. The red line locates the cross-section in Figure 7.5 and 7.6 and the red box delineates the area in perspective view in Figure 7.6

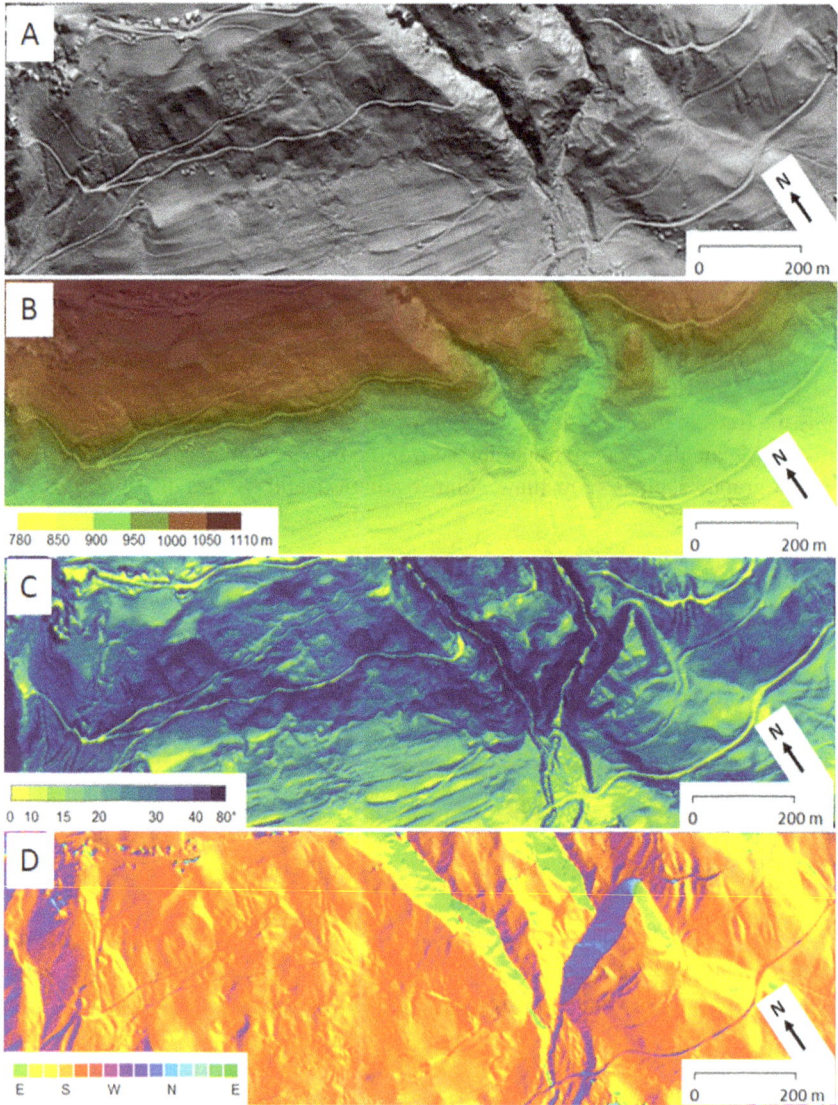

FIGURE 7.8   Digital terrain model derived as a grid of 0.2 metre cell size. (A) shaded relief, (B) elevation combined with shaded relief and using a grid of 2 metre cells (C) slope angle, and (D) slope aspect.

One of the main benefits of mapping forests by ULS is the calculation of forest metrics with the ability to locate and measure individual trees. Figure 7.9 shows the results of the procedure *segment_trees* using the lidR routine in the software package R (Roussel et al., 2020). Location of the area is marked by the red square in Figure 7.7. This high data density and accuracy could be used to calculate ground biomass,

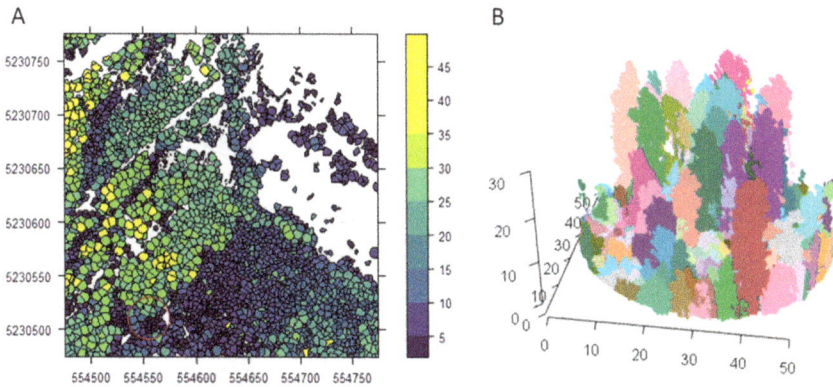

FIGURE 7.9 (A) Polygons delineating segmented trees from the ULS point cloud coloured by maximum tree height in meters. (B) A 3D perspective view of a group of segmented trees with unique tree colours from the area marked with red circle in A. The units of axes coordinates are in meters.

estimate log production volumes, and aid forest management plans (Bruggisser et al., 2020).

MODELLING SOLAR IRRADIATION ON TREES

Contemporary solar modelling tools do not allow for modelling the solar irradiation simultaneously for a ground surface and tress.

To calculate this, the program v3.sun, which is designed for modelling solar energy in high spatial resolution across areas of several square kilometers on complex 3D objects was used (Kaňuk et al., 2019). The workflow for this program is shown in Figure 7.10. This program is run via a series of steps using GRASS GIS Shell Scripts and uses data structures derived by adaptive triangulation from 3D point clouds.

3D models of the area of interest (Figure 7.11 (A)) were derived from the ULS point cloud using 3D Delaunay triangulation applied in Geomagic Wrap software (Edelsbrunner et al., 1998). This is an interpolation method based on an adaptive triangulation. On the triangular irregular network (TIN) surface produced from this modelling, it is necessary to orient the normals of the surface for each triangle to the sun (Figure 7.11 (C)). Determining this orientation of triangle normals plays a key role in preparing the input data for modelling solar irradiation.

The v3.sun module calculates the irradiance/irradiation value for a user-defined time interval (Figure 7.10 (F1)), and the position of the sun is determined according to a defined step within the astronomical day (Figure 7.10 (F2)). Subsequently a tracking solar beam for each face of each triangle is produced. The result is stored in the 'energy' parameter as $W.m^{-2}$ (watts per square meter). Thus, it is the amount of energy that has impacted on the area of the face of triangle.

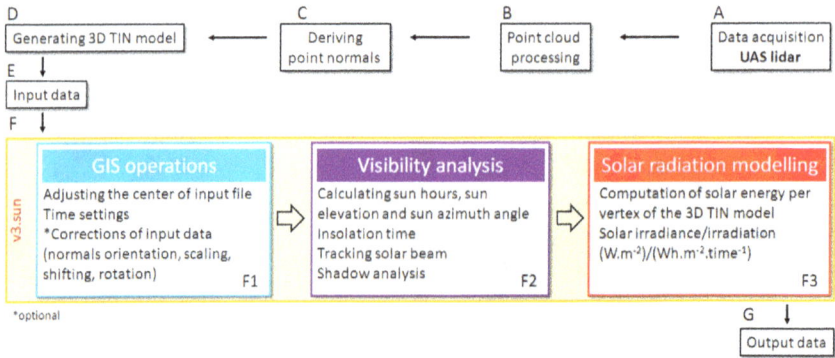

FIGURE 7.10 Workflow of solar radiation modelling on the 3D forest mesh by the v3.sun module prototype.

Source: Kaňuk et al., 2019

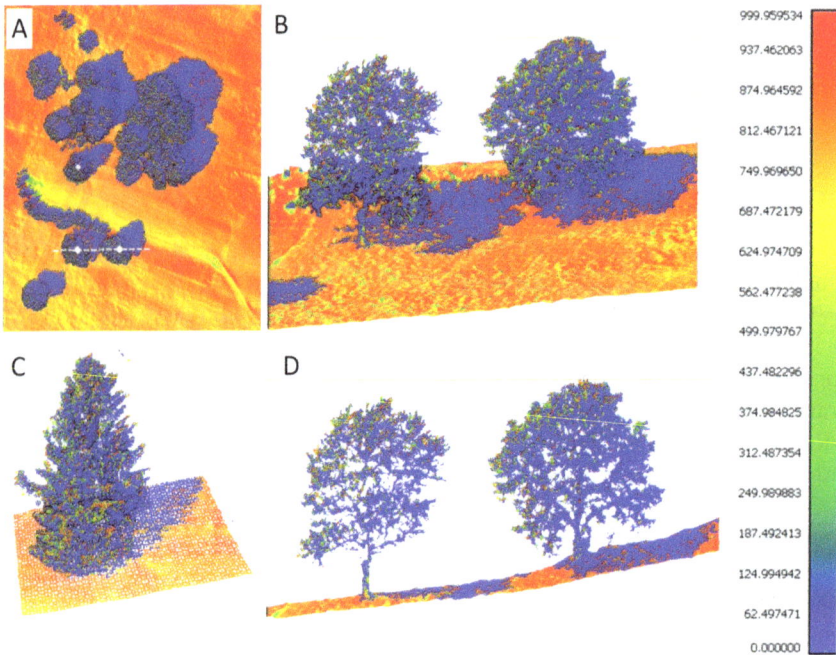

FIGURE 7.11 3D mesh from the ULS data coloured by direct solar irradiation (W.m^{-2}) on 21 June 2017, 12:00 local time, generated using v3.sun. Top orthogonal view (A) shows the location of the tree tops displayed in the 3D views of (B) and (C). The dashed line in (A) locates the vertical cross-section in (D).

The advantage of this approach is a highly detailed model of solar energy incident to the tree and the surface below, regardless of the geometric complexity of the tree. The 3D mesh of the 100 m x 100 m selected area and the results from this process are shown in Figure 7.11. This scene contains 6 million facets/triangles. It is the authors' intention that the v3.sun program will become an open-access tool for highly detailed solar radiation modelling for geometrically complex 3D landscape objects such as forests or urban landscapes.

CONCLUSIONS

This case study demonstrated that ULS enables the collection of ultra-high-density point clouds using wider laser scan angles than have been previously possible from traditional higher altitude airborne platforms. These low-altitude drone flights make it possible to acquire 3D measurements with high precision and accuracy, achieving point densities of thousands of points.m^{-2}. On forests, this data can be used to clearly resolve branch and stem structures, comparable to TLS and is acquired more rapidly over large landscapes at a fraction of the cost of traditional airborne laser scanning. Dense 3D point clouds, capturing the complexity of the vertical tree profiles, open opportunities to model complex natural phenomena such as solar energy. Unpiloted laser scanning is not a replacement for piloted airborne platforms or TLS. ULS represents a binding link between ALS and TLS, providing point clouds with levels of detail close to TLS, while facilitating a more time-efficient acquisition of larger areas, amounting typically to several hectares. Drone-based flight operations provide flexibility and enable access to locations where traditional flight operations are challenging, either because the sites are remote or because permissions are difficult to secure. The market of lightweight LiDAR systems and drone platforms is growing rapidly providing wide range of technological solutions. Further reduction in size and weight with an increase in device performance paves a very promising future for smaller and lighter ULS systems.

ACKNOWLEDGEMENTS

The presented research originated thanks to the financial support of the Ministry of Education, Science, Research and Sport of the Slovak Republic under the grant nr. VEGA 1/0085/23 'Modeling urban heat islands using geospatial tools'.

REFERENCES

Brede, B., Calders, K., Lau, A., Raumonen, P., Bartholomeus, H.M., Herold, M., Kooistra, L. 2019. Non-destructive tree volume estimation through quantitative structure modelling: Comparing UAV laser scanning with terrestrial lidar. *Remote Sensing of Environment* 233, Article 111355.

Bruggisser, M., Hollaus, M., Kükenbrink, D., Pfeifer, N. 2019. Comparison of forest structure metrics derived from UAV lidar and ALS data. *ISPRS Annals of the Photogrammetry, Remote Sensing and Spatial Information Sciences* 4(2/W5), 325–332.

Bruggisser, M., Hollaus, M., Otepka, J., Pfeifer, N. 2020. Influence of ULS acquisition characteristics on tree stem parameter estimation. *ISPRS Journal of Photogrammetry and Remote Sensing* 168(2020), 28–40.

Edelsbrunner, H., Facello, M.A., Fu, P., Qian, J., Nekhayev, D.V. 1998. Wrapping 3D scanning data. In *Three-Dimensional Image Capture and Applications* (Vol. 3313, pp. 148–159). International Society for Optics and Photonics.

Gallay, M., Eck, C., Zgraggen, C., Kaňuk, J., Dvorný, E. 2016. High resolution airborne laser scanning and hyperspectral imaging with a small UAV platform. *The International Archives of the Photogrammetry, Remote Sensing and Spatial Information Sciences* XLI-B1, 823–827.

Hu, T., Sun, X., Su, Y., Guan, H., Sun, Q., Kelly, M., Guo, Q. 2021. Development and performance evaluation of a very low-cost UAV-lidar system for forestry applications. *Remote Sensing* 13(1), 77.

Hyyppä, E., Hyyppä, J., Hakala, T., Kukko, A., Wulder, M.A., White, J.C., Pyörälä, J., Yu, X., Wang, Y., Virtanen, J.-P., Pohjavirta, O., Liang, X., Holopainen, M., Kaartinen, H. 2020. Under-canopy UAV laser scanning for accurate forest field measurements. *ISPRS Journal of Photogrammetry and Remote Sensing* 164, 41–60.

Kaňuk, J., Gallay, M., Eck, C., Zgraggen, C., Dvorný, E. 2018. Technical report: Unmanned helicopter solution for survey-grade lidar and hyperspectral mapping. *Pure and Applied Geophysics* 175(9), 3357–3373.

Kaňuk, J., Zubal, S., Šupinský, J., Šašak, J., Bombara, M., Sedlák, V., Gallay, M., Hofierka, J., Onačillová, K. 2019. Testing of V3.sun module prototype for solar radiation modelling on 3D objects with complex geometric structure. *International Archives of the Photogrammetry, Remote Sensing and Spatial Information Sciences—ISPRS Archives* 42(4/W15), 35–40.

Kellner, J.R., Armston, J., Birrer, M., Cushman, K.C., Duncanson, L., Eck, C., Falleger, C., Imbach, B., Král, K., Krůček, M., Trochta, J., Vrška, T., Zgraggen, C. 2019. New opportunities for forest remote sensing through ultra-high-density drone lidar. *Surveys in Geophysics* 40(4), 959–977.

Lambert, J., Carballo, A., Cano, A.M., Narksri, P., Wong, D., Takeuchi, E., Takeda, K. 2020. Performance analysis of 10 models of 3D LiDARs for automated driving. *IEEE Access* 8, Art. no. 9142208, 131699–131722.

Morsdorf, F., Eck, C., Zgraggen, C., Imbach, B., Schneider, F.D., Kükenbrink, D. 2017. UAV-based LiDAR acquisition for the derivation of high-resolution forest and ground information. *Leading Edge* 36(7), 566–570.

Roussel, J.R., Auty, D., Coops, N.C., Tompalski, P., Goodbody, T.R.H., Sánchez Meador, A., Bourdon, J.F., De Boissieu, F., Achim, A. 2020. lidR: An R package for analysis of Airborne Laser Scanning (ALS) data. *Remote Sensing of Environment* 251, 112061.

Wang, Y., Pyörälä, J., Liang, X., Lehtomäki, M., Kukko, A., Yu, X., Kaartinen, H., Hyyppä, J. 2019. In situ biomass estimation at tree and plot levels: What did data record and what did algorithms derive from terrestrial and aerial point clouds in boreal forest. *Remote Sensing of Environment* 232, 111309.

Wieser, M., Mandlburger, G., Hollaus, M., Otepka, J., Glira, P., Pfeifer, N. 2017. A case study of UAS-borne laser scanning for measurement of tree stem diameter. *Remote Sensing* 9(11), 1154.

8 Digital Mapping and Recording of Inishtrahull Island and Its Built Heritage in 24 Hours

John Meneely, Kendrew Colhoun, Trevor Fisher,
Michael Casey, Daniel Moloney, and Alan Lauder

HISTORY OF THE ISLAND

The island of Inishtrahull, home to Ireland's northernmost lighthouse, lies approximately 10 km northeast of Malin Head on the coast of Donegal, Ireland (Figure 8.1). The island is formed of granitic gneiss, and this metamorphic rock, dated radiometrically by Daly et al. (1991) to 1,779 ± 3 million years, makes them Ireland's oldest rocks. The first evidence of human presence comes from middens found on the island that contained flints. These flints date from the early Mesolithic period (at least 5,000 years ago). It is not known with certainty how long the island was inhabited, but it appears that there was a long period of time with no permanent human habitation prior to 1700. After 1700, a small community became established, reaching a peak population of approximately 80 people towards the end of the 19th century. The island still had a small resident population until 1929 and the lighthouse was manned until 1987. Today it is uninhabited and has been designated a protected area due to its wildlife. The island has many derelict dwellings and larger structures dotted over its surface.

THE FAMILIES OF INISHTRAHULL

Inishtrahull, for its size, had a remarkably large resident population from the mid-1800s through to 1928, when the island was evacuated. All the houses were in the relatively flat centre of the island, known as the Laggan (red oval in Figure 8.3). This area, between the high ground to the east and west, was the most sheltered region on the island and had most of the fertile ground, for growing crops and grazing livestock.

Irish census returns from 1901 and 1911 were used to gather information on who lived on the island. In the 1901 census, there were 65 people living there, including lighthouse keepers in 13 separate houses. There was also a separate census return form for each house, plus an 'Enumerator's Abstract', which listed all the houses; a

DOI: 10.1201/9780429327575-8

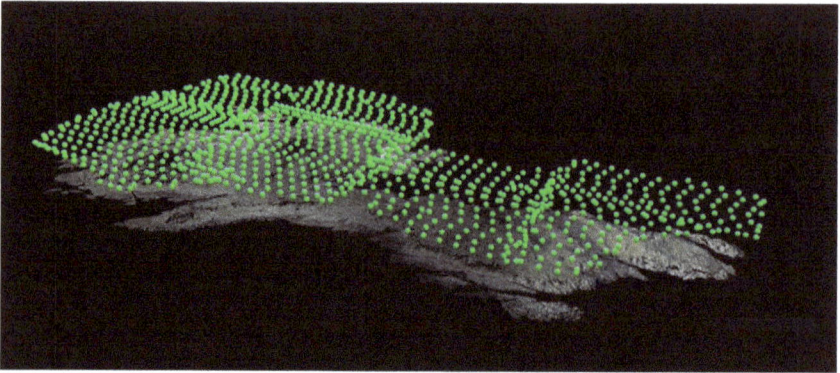

FIGURE 8.1   3D location of drone images—green spheres—used to create a 3D map of Inishtrahull Island.

'House and Buildings' return, which provides an overview of all houses and families and finally; and an 'Out Houses and Farm Steadings' return, which gives an outline of additional outhouses attached to each residence. In the 1911 census, 81 people lived in the island's 21 occupied buildings. These buildings were roofed with either thatch (15 houses) or slates/corrugated steel (6 houses), and these are mainly associated with the lighthouse.

The relatively restricted Laggan area of the island was used to supplement the main activities of the islanders rather than the main source of their livelihood. The income of the island's 21 households in 1911 was dominated by two activities: fishing and servicing the lighthouses and communication infrastructure. Then, as now, that infrastructure visibly dominated the island. In 1911 that infrastructure was a crucial element of the island's economy. In 1911, 13 of the 21 occupied houses identified fishing as the principal source of their livelihood, while other households derived their living from Lloyd's of London or the lighthouse, fog signal, and Marconi Relay Station. The remaining households were those of the schoolteacher and one house whose head stated her main occupation was 'farmer and boat contractor'. As well as those directly employed as signalmen and lighthouse keepers, the lighthouse and associated signalling infrastructure would have generated a demand for casual labour, boat and ferry services, and the delivery of supplies including food and fuel.

DATA COLLECTION AND PROCESSING

To create a high-resolution 3D digital base map of the whole island, a Yuneec H520 drone was used to collect 1,032 separate images from approximately 90 m above its surface (Figure 8.1) These images were collected in 9 separate flight missions of approximately 25–30 minutes each. Prior to these flights, 15 ground control points (GCPs) were marked on the ground in the form of a large white numbered cross at strategic intervals over the surface of the island with survey spray paint. There are a number of basic requirements for GCPs. The main one is good visibility so that they can be found in drone images. Typically, a GCP should be approximately half a metre by half a metre (2 ft x 2 ft)—and painted in highly contrasting colours to differentiate it from its surroundings and, its centre point should also be easily identified. In this case, a small stone that was sprayed white was placed in the centre of each cross to improve identification of this point where the numbered cross was sprayed on rough ground. The number of GCPs you require depends on the size and topology of your survey site. It is important that all GCPs are evenly allocated through your survey area.

The location of these GCPs was recorded in Irish Transverse Mercator (ITM) using a Leica survey-grade GNSS system. The coordinates of these GCPs were then used by the drone mapping software to georeference the data—accurately position the map in relation to the real world around it.

The 1,032 images and GCP location data were processed using Pix4D™, a specialist survey drone photogrammetry software to produce a 3D model of the island in the ITM coordinate system. Additional output from this processing included a 2D orthomosaic image of the entire island with a resolution of approximately 30 mm per pixel (Figure 8.2), a 3D 1 m contour file in .dxf format (Figure 8.3), a digital terrain map (DTM), and a digital surface model (DSM)—a 2.5D model of the mapped area that contains (x,y,z) information but no colour information for use in GIS software (Figure 8.4).

It should be noted that processing such large data sets using this or other specialist drone mapping software on a local computer requires a high-end graphics card and at least 32 GB of RAM. This data set was processed on an i9i7 laptop with 64 GB of RAM and a 2010 NVidia graphics card, and it still took over 24 hours to process at its most detailed settings. Some drone mapping software companies offer a cloud-based processing option, and this may be a more cost-effective solution for some than purchasing a high-spec computer.

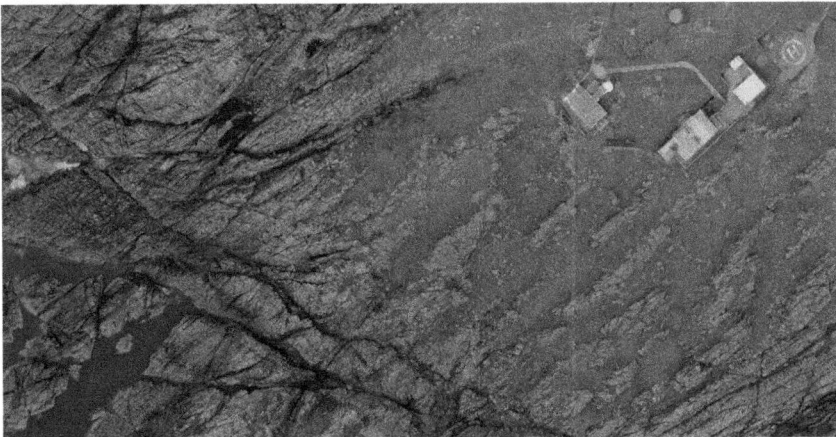

FIGURE 8.2 A section of the 2D orthomosaic image showing the helipad, the old light-house keeper dwelling, and the new lighthouse. The resolution on this image is approximately 30 mm per pixel.

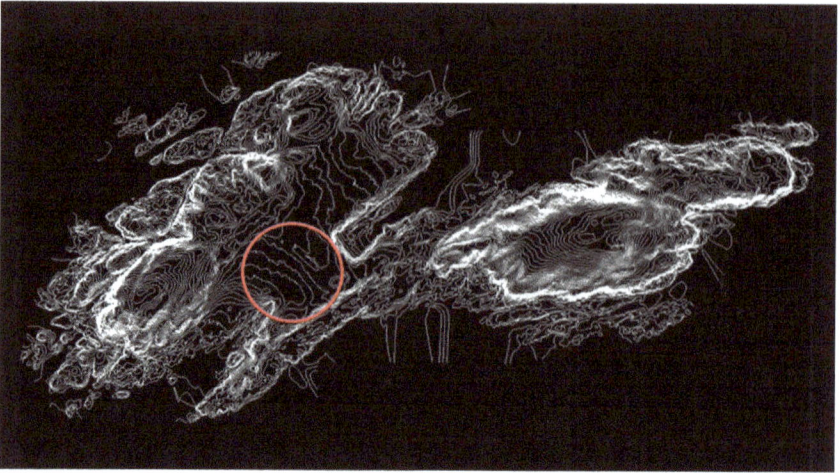

FIGURE 8.3 A 1 m contour map of Inishtrahull Island, generated in Pix4D from drone imagery. The relatively flat area highlighted in red is known as the Laggan.

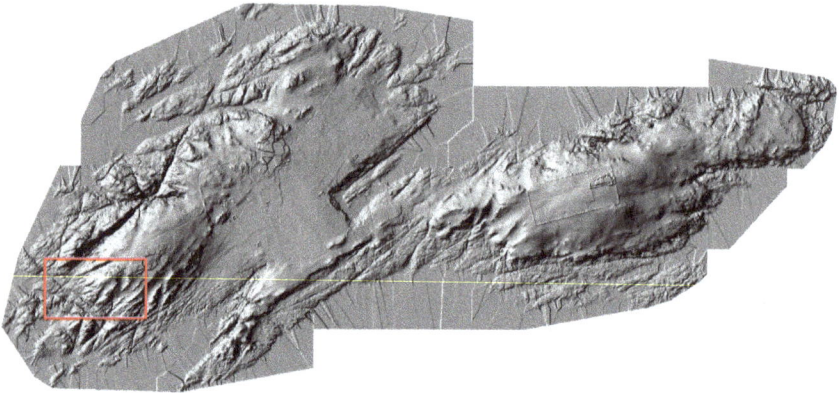

FIGURE 8.4 A 2.5D Hillshade image of the whole island using the DSM generated by Pix4D in QGIS. The area highlighted in red is the location of Figure 8.2.

In addition to the all-island 3D model, additional drone imagery was gathered to produce higher-resolution 3D models for four of the island's main buildings (Figure 8.5). These images were collected using a DJI Mavic Pro 2 drone from a height of approximately 20 m above each structure. Automated flight paths and overlapping image collection on these buildings were controlled using the free flight planning and execution app Pix4D Capture. Other free mapping apps are available, and most are extremely easy to use and can be run on a smartphone or tablet attached to the drone's remote control unit. They usually require the user to first decide if the imagery to be collected is for the production of a 2D orthophoto or 3D model, then define the area to be surveyed by constructing a polygon on a background map or satellite

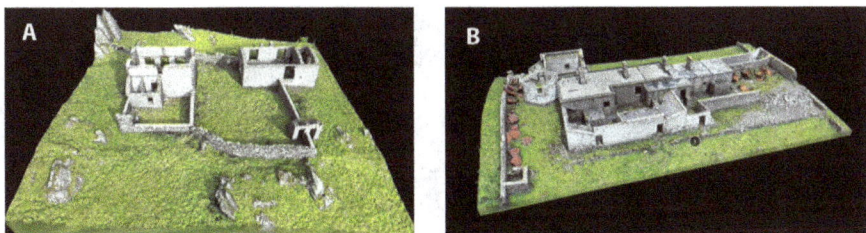

FIGURE 8.5   3D models of (A) the School House and (B) The Old Lighthouse.

image and input a height the mission is to be flown. If necessary the user can define the camera angle and image overlap. The software will then calculate how long the survey should take and what the ground sampling distance (GSD) will be—this is the distance between two consecutive pixel centres measured on the ground. The bigger this value, the lower the spatial resolution of the image and the less visible details will be.

Once 'mission start' is selected, the drone takes off, flies the survey mission, and returns to its departure location completely autonomously. During the flight, the drone operator must keep alert, continually watch the drone and the surrounding area, and always be ready to take over its control manually should it be necessary. It is important to note that drone legislation varies from country to country so always be aware of the rules that govern the area to be flown.

In total 312 images were collected over the Lighthouse, 285 images over the Old Lighthouse, 340 images over the School House, and 224 images over the Graveyard. These images were then processed using software from Reality Capture (RC), a structure from motion (SfM) software package, to produce detailed 3D models of each site (Figure 8.5). SfM is a photogrammetric, range-imaging technique for generating three-dimensional models from two-dimensional image sequences. This SfM software was also used to generate 3D models of a Celtic Cross, shown in Figure 8.6, and one of the remaining gravestones in the graveyard at the eastern end of the island. Multiple, 12 MB, overlapping images on these two objects were taken with a Sony A7R IV digital camera and processed with RC. These models were uploaded to Sketchfab and links were created to them from the online 3D model of the graveyard, which has a link from the 3D model of the whole island. All these SfM models can be found on Sketchfab at https://sketchfab.com/1manscan.

In addition to the drone mapping, ultra-high-resolution 3D terrestrial laser scanning (TLS) was carried out on the internal and external structure of the Old Lighthouse, the ruins of Daniel Hutton's House, and the external of the new Lighthouse. As mentioned previously, a drone survey of the Old Lighthouse was also carried out (Figure 8.5), but as this was an aerial survey and contained no information on the internal layout of this structure. However, it did contain information on the roof of the building—data that was not obtained by the TLS.

In total 117 individual 360° 3D laser scans were collected across all three sites using a FARO Focus™ S150 tripod-mounted 3D laser scanner—resulting in 1.3 billion individual measurements collected across all three structures. Because of the

FIGURE 8.6   3D model of Celtic cross produced using SfM on Inishtrahull.

sheer amount of measured information collected by 3D laser scanners, these data sets are also known as point clouds. An individual measured point in a point cloud is stored as an x,y,z coordinate and an intensity value (I). A somewhat simplified definition of this intensity value, also known as a reflectance value, is how much of the laser light that is sent out to a surface did the instrument receive back. There are several factors that affect intensity including how much light was sent out, the range to and reflectance of the target, attenuation by the atmosphere, and the incident angle of the light on the target surface. The 3D laser scanner can also collect colour imagery of the area measured and map this onto the 3D laser measurements as a RBG value. On this TLS system, the colour imagery is collected after the measurement phase and adds an additional three minutes to each laser scan. Due to the time constraint to carry out all the 3D surveys on the island, this feature was not enabled.

The individual scans from each site were pre-processed and then registered (stitched together) in FARO SCENE software using a process known as 'cloud-to-cloud' (C2C) to form one large, combined point cloud of the structure. Pre-processing is an automated process that can improve the quality of the data and reduce the probability of outliers or noise interfering with data analysis while maintaining the clarity of features. This is achieved using a number of user-defined filters. These include: (a) a dark scan point filter, which removes scan points below a certain reflectance value as dark surfaces absorb light, and if there is not enough light returning to the scanner, scan points can be less accurate; (b) a distance filter, which deletes points outside of a specified distance range set by the user; and (c) a stray point filter, which compares each point's distance to the points around it to determine if the point should be included in scan data.

There are several methods that can be used to register individual laser scans, including manual placement (low accuracy) to the positioning of reference targets in a scan. These targets (usually a black-and-white, printed flat quadrat approximately 30 cm square or a white 20 cm diameter sphere) are placed in the area to be scanned and must not be moved between successive scans as they act as reference points which are automatically detected by the registration software when aligning these scans (high accuracy). The placement and movement of a minimum of three of these targets between every other scan is time-consuming, so for this reason, this method was not employed and the scans were registered by C2C. This registration method uses automatically extracted features and scan points in the overlap area of scans for alignment. The advantage of this form of registration is that in some cases it may be more accurate than target-based, which can use a minimum of only three points at a time. Also, targets may move, a checkerboard may dislodge, or a sphere be accidentally touched. The disadvantage of C2C is that you must have a decent overlap between scans. C2C's main benefits are that it is targetless and the process is fully automated. However, this means that hundreds of thousands of scan points may be used, which can lead to longer post-processing times. Unlike target-based, some environments with repeating geometries like tunnels and roads as well as environments with little geometry, like the surface of a field, will not work with C2C. One thing to note is that in larger projects (greater than 500 scans), it is customary to use a combination of both target-based and C2C. Whatever registration is chosen to align individual scans, it is critical that the final registered data is reviewed to guarantee close alignment to real-world conditions.

The data from these TLS sites can be exported in a wide variety of formats and imported into most modern GIS or 2D/3D CAD packages to create 2D plans, elevations, cross-sections, DTMs, and millimetre-accurate 3D models. This is an unparalleled 3D digital record of these structures. Figure 8.7 shows a 50 cm horizontal slice through the point cloud collected on the Old Lighthouse. This was generated by exporting the registered 3D laser scan data as a ReCAP (Reality Capture) file. This is a format used by Autodesk software to store spatially indexed point cloud data. It was then imported into Autodesk AutoCAD 2023 for slicing and annotating with a scale bar and north arrow. The point cloud can also be sliced vertically to create cross-sections anywhere throughout the structure (Figure 8.7).

To visualise and share this wealth of census information on who occupied these now derelict dwellings and their location on the island, a 3D model generated from the whole island drone survey was uploaded to Sketchfab.[1] This is a platform to publish and share 3D models. It provides a viewer based on WebGL and WebXR technologies that allow users to display 3D models on the web. The 3D models uploaded to this platform can be embedded in websites or social media posts to be viewed on any mobile browser, desktop browser, or Virtual Reality headset. Two identical 3D models of the island were uploaded, and the locations of all built heritage structures were annotated on one model and wildlife and places of interest on the other model—this annotation process is performed on the Sketchfab website prior to publication. When clicked on, these annotations on both models expand to show images, additional material, and links to the IBO website where there is more detailed information on who lived there, when they lived there, and images of the occupants. Some annotations on the built heritage model have links to more detailed 3D models of that structure, which are stored on the same site (Figure 8.8).

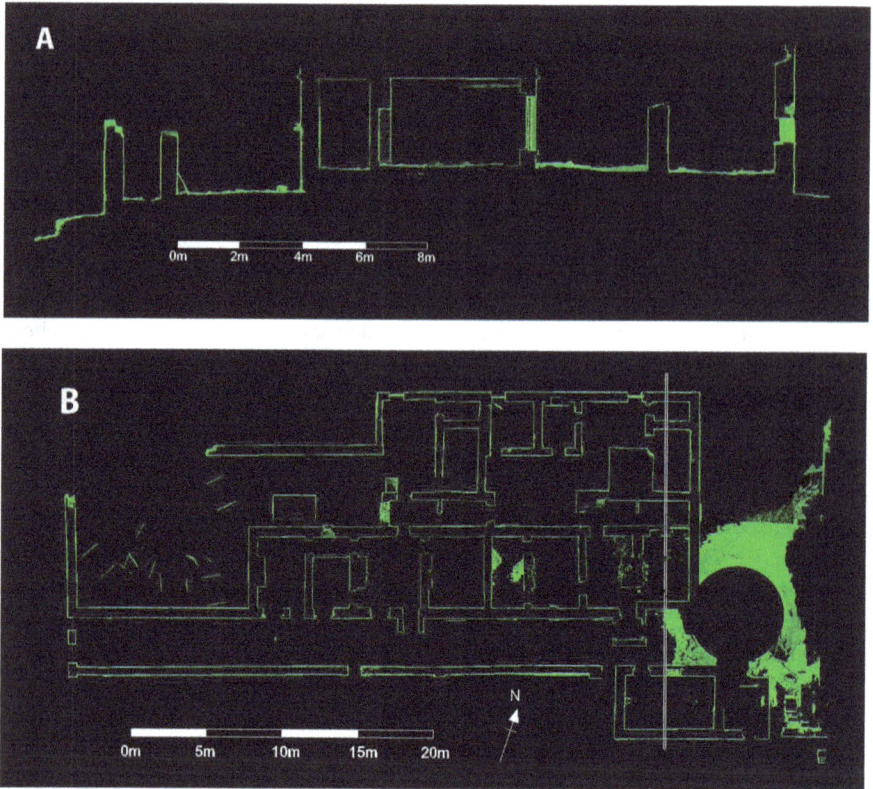

FIGURE 8.7   (A) A 50 cm horizontal slice through the laser scan data collected on the Old Lighthouse. (B) A 50 cm vertical slice through the lighthouse point cloud—white line in top image.

These online annotated 3D models of the island are very useful in demonstrating its overall geomorphology and the relationship between this and its built heritage. They are also an excellent online resource to show the history of the island's occupation and allow virtual visits for those unable to travel to this very remote location.

Most major 3D laser scanner manufacturers offer a cloud-based service to host and share processed 3D laser scan data across the Internet (e.g., FARO Webshare™ and Leica TruView ™), and there are a large number of independent suppliers of this service. These online facilities let the user view and interrogate the data in 2D plan view (Overview Map) or move from individual 3D scan to 3D scan (Scan View) within a project, much like Google Street View, with the added benefit of being able to measure areas and distances and annotate the scan data (Figures 8.9 and 8.10).

Figure 8.9 shows the Overview Map function on this platform. The individual scan locations are displayed as a Map Pin, colour-coded by height and labelled Inish_001 etc, and these can be clearly seen around the perimeter of the building. The coloured circles are the locations of clusters of scans, with the number of scans in that cluster at its centre. Their individual location is not discernible at this zoom level. By

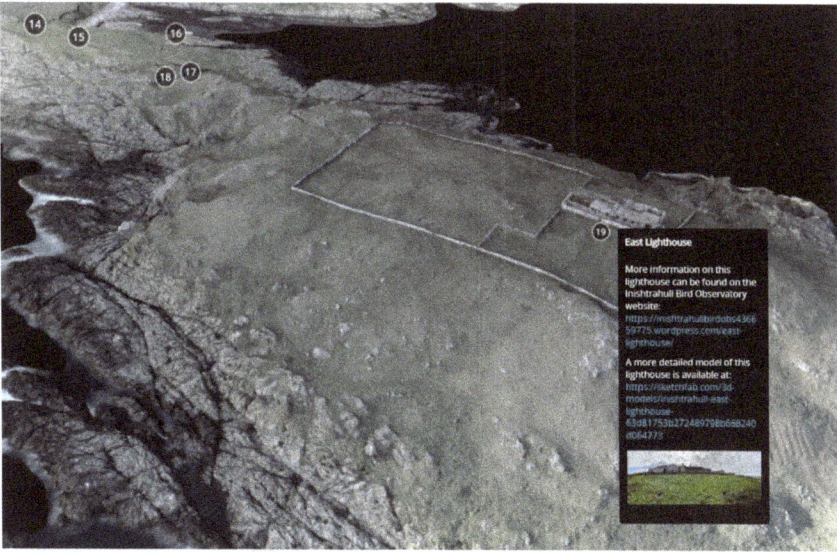

FIGURE 8.8 This image shows the Built Heritage 3D model of Inishtrahull Island with annotations/locations. The 'pop-out' window shown, which appears when a numbered location is clicked, provides a link to the IBO website where more information on the OldLighthouse can be found. It also provides a link to a more detailed 3D model and a thumbnail image of this structure.

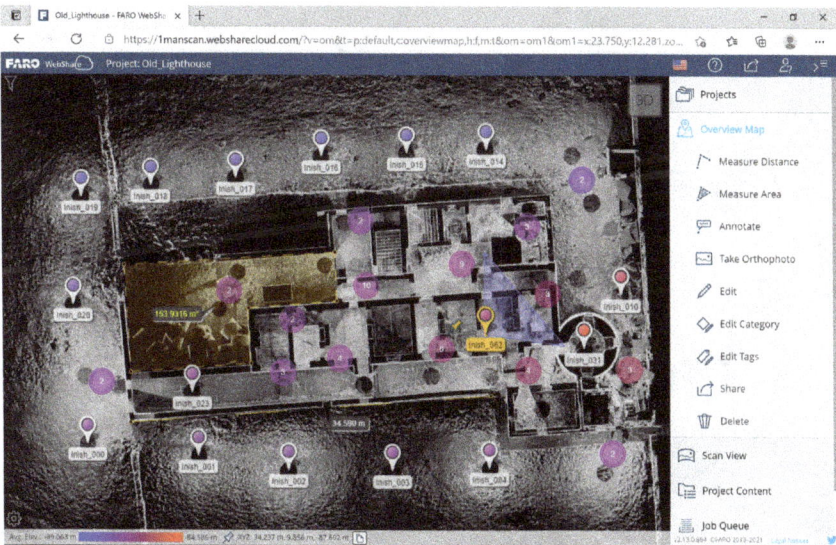

FIGURE 8.9 This is an image of the overview map of the Old Lighthouse on the Webshare portal.

selecting various functions from the menu on the right of this image, measurements, annotations, and orthophotos can be made from this overview map. The orange region on the left of this image is an area measurement. By simply tracing the boundary on screen with the Measure Area function, an area is calculated—153.9316 m². There is also a distance measurement along the front of the building in this image of 34.500 m—created by selecting the Measure Distance function and clicking on two points on the overview map.

Figure 8.10 is a scan view image of the file Inish_062. To enter this, simply double-click on any of the scan location map pins on the overview map. This takes you into a spherical 360° view of that scan. As in the overview map, distances can be measured and/or annotations can be left in the scene. This greyscale image is generated from the reflectance values collected during the scan. If colour imagery had been collected, this can also be viewed or a mixture of both by selecting the colour wheel icon in the bottom left of the screen. The measurement in this image (blue-black line) was made to calculate the ceiling height in this room. One point was selected on the floor and one on the ceiling—these do not have to be directly above each other. As each pixel in this image also has an x,y,z value, the software can calculate the actual distance between these two points—2.782 m, the horizontal offset between them—0.633 m and the corrected vertical distance—2.709 m. The map pins in this figure are the locations of other scans in that area—the larger the pin, the closer it is to the current location. To 'jump' to that location, simply double-click on it. If at any stage in a scan view, you want to check its location, simply return to the overview map and the scan is highlighted in orange with a blue triangle showing the current direction and field of view (Figure 8.9)

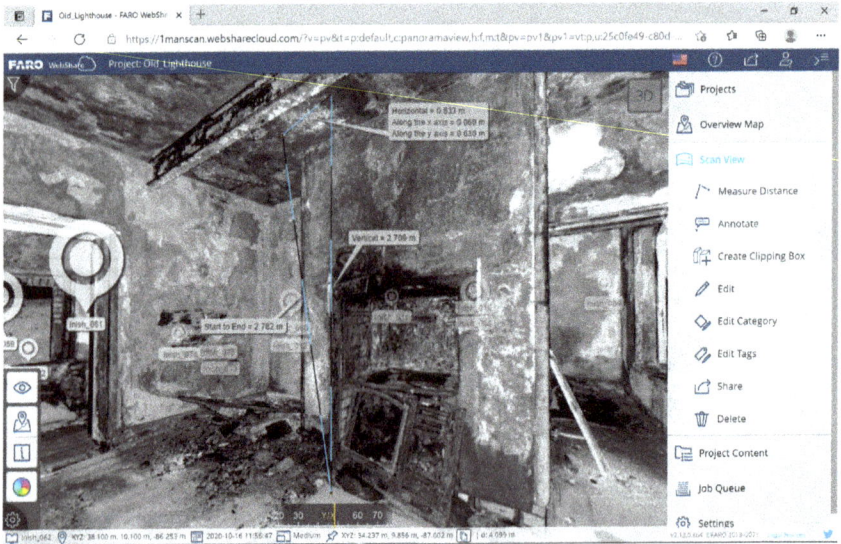

FIGURE 8.10   A scan view image of the laser scan file Inish_062—highlighted in orange in Figure 8.9.

The ability to share/visualise this or any laser scan data via a web portal and interrogate and/or annotate it makes it an exceptionally powerful tool. These platforms give secure, cloud-based global project/facility management capabilities so every stakeholder can remain informed with web access to accurate 3D reality data. They can also be used to continuously monitor as-built data during construction or to integrate 3D reality data as part of a digital twin.

CONCLUSION

It is clear from this chapter that a large amount of 3D digital data can be collected in a relatively short time by a small group of people. Recent advances in these 3D technologies, especially in the areas of speed of data collection, portability of the equipment, and software processing allowed this team to capture a whole island and most of its built heritage in 24 hours (2 days in the field). Innovations in how this digital data can be visualised, shared, and interrogated online, especially when it can be linked to other data sources, in this case census data, bring it to life. This digital data will also be used to help frame and inform forthcoming conservation plans for the island, create baseline data for future research, and create engaging methods to share all this information online.

NOTE

1. https://sketchfab.com/1manscan

REFERENCE

Daly, J.S., Muir, R.J., Cliff, R.A.A. (1991) A precise U-Pb zircon age for the Inishtrahull syenitic gneiss, County Donegal, Ireland. *Journal of the Geological Society* 148(4): 639–642.

9 CHERISH

Development of a Toolkit for the 3D Documentation and Analysis of the Marine and Coastal Historic Environment

Anthony Corns, Robert Shaw, Linda Shine, Sandra Henry, Edward Pollard, Toby Driver, Louise Barker, Daniel Hunt, Sarah Davies, Patrick Robson, Hywel Griffiths, James Barry, Kieran Craven, and Sean Cullen

INTRODUCTION

CHERISH is a EU-funded Ireland-Wales project that aims to raise awareness and the understanding of the past, present, and near-future impacts of climate change on the rich cultural heritage of the Irish and Welsh regional seas and coast. The project began in January 2017 and will run until June 2023 and will benefit from €4.9 million of EU funds through the Ireland Wales Co-operation Programme 2014–2020.

The project is targeting work in several studies, 5 areas of Ireland from Co. Meath on the east coast to Co. Kerry on the west coast, and 13 areas in Wales covering Anglesey, Gwynedd, Ceredigion, and Pembrokeshire (Figure 9.1). These study areas encompass both individual sites and wider landscapes, from promontory forts and shipwrecks through to islands and dune systems, including a range of environments from marine inshore waters into the intertidal zone and onto the coast edge.

PHILOSOPHY OF THE TOOLKIT

During project planning, it was realised that a complementary range of approaches and methods used across research domains that include archaeology, geography, and geology was required to better understand, record, and monitor the coastal and marine historic environment. Following input from all project partners and external agencies, the CHERISH toolkit was developed to provide an integrated approach to coastal and marine recording (see Figure 9.2). Several factors influenced the selection of appropriate methods. This chapter explores some of the methods employed to digitally document the coastal and marine historic environment in 3D. Other

DOI: 10.1201/9780429327575-9

FIGURE 9.1 Illustration identifying the 5 Irish and 13 Welsh study areas within the CHERISH Project and some of the selected sites where the toolkit is being employed.

FIGURE 9.2 Illustration of the different survey approaches employed by CHERISH to record the coastal and marine environment 1. Airborne laser scanning (ALS), 2. Unmanned aerial vehicle (UAV or drone) survey 3. Satellite mapping, 4. Aerial survey, 5. Geophysical survey, 6. Coring, 7. Precision survey, 8. Erosion monitoring, 9. Terrestrial laser scanning, 10. Excavation, sampling, and dating, 11. Marine mapping, 12. Underwater archaeological survey.

components employed in this research such as peat coring and sampling are not covered in detail here.

WORKING IN A DYNAMIC ENVIRONMENT

Effective and accurate surveying and recording in the marine and coastal environment can be difficult due to the practicalities of access, tides, and the exposed environment (Kotilainen and Kaskela 2017). The nearshore marine and coastal environment are often referred to as the 'white ribbon' (Leon et al. 2013) due to the highly dynamic environment between high and low water, creating challenges for the efficient collection and integration of elevation data (Driver and Hunt 2018). Using a combination of methods and techniques, the CHERISH toolkit approach enables this zone to be effectively recorded. Remote aerial technologies such as ALS and UAV surveying allow accurate recording with reduced risk to the surveyors. It can be more challenging to effectively record wrecks and other structures beneath the sea. Selecting methods that can be safe and relatively easily applied underwater and on the surface increases the opportunities for successful data collection—these include remotely operated underwater vehicles (ROVs) and multi-beam bathymetry.

SCALE, ACCURACY, RESOLUTION, AND EFFICIENCY

When selecting relevant geomatic techniques, the ability to identify, detect, and monitor change in cultural and environmental features must be considered (Guisado-Pintado et al. 2019). The scale and nature of cultural features and coastal change which may occur will influence which technique is appropriate (Boehler et al. 2001) (Figure 9.3). The ability to record high-resolution and accurate data must also be offset against the efficiency of the proposed method and the overall size of the survey area. Employing an approach such as terrestrial laser scanning (TLS) may produce the most accurate method of recording a coastline. However, the time and cost required to employ such a method on a regional scale would be prohibitively expensive. In addition, when selecting appropriate methods, consideration must be given to the temporal resolution which can be achieved with the chosen method, particularly with respect to the frequency of repeat survey and the relative ease and costs.

DATA INTEGRATION, PRESENTATION, AND REUSE

In the collection of 3D data for the marine and coastal environment, techniques have been selected which offer the best integration of data sets into several methodological processes which enable the maximum amount of research return from their collection. Standardised data outputs such as point clouds, digital elevation models (DEMs), and orthoimagery can be utilised with geoprocessing analysis to inform experts on a range of features and change detections, including the identification of new archaeological sites to the monitoring of erosion of submerged wrecks. This data can also be repurposed into highly informative and engaging visualisations to enable experts to advise and educate wider non-scientific stakeholders about the

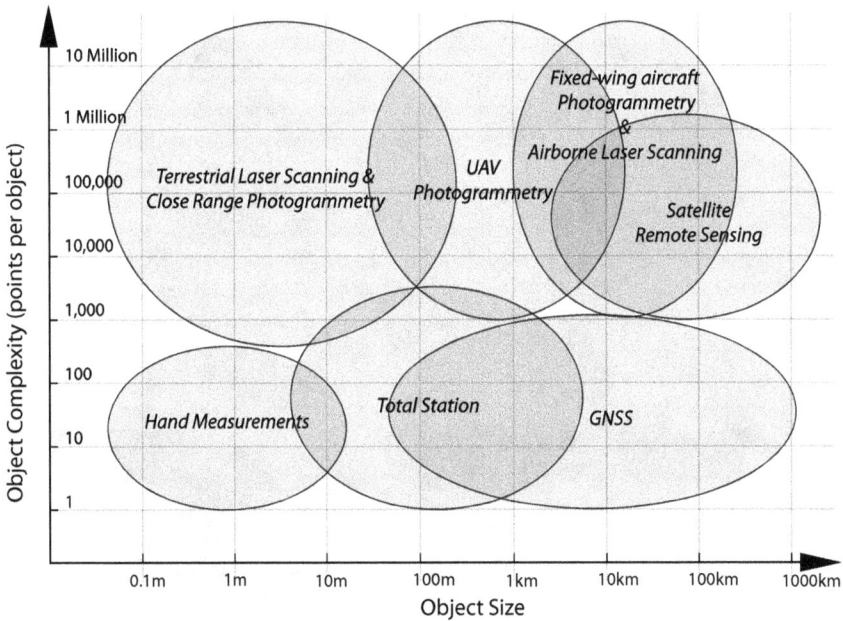

FIGURE 9.3 Illustration of the relationship between appropriate survey techniques for the scale of objects and the relative complexity of recording (after Boehler 2001).

challenges faced in mitigating and managing the effects of climate change on coastal and historic marine environments.

TERRESTRIAL MAPPING AND MONITORING OF COASTAL SITES AND LANDSCAPES

ESTABLISHING MONITORING NETWORKS—USING GLOBAL NAVIGATION SATELLITE SYSTEMS (GNSS) IN THE COASTAL ZONE

At the heart of any recording/monitoring project is the need to collect highly accurate and precise measurements. Using Global Navigation Satellite Systems (GNSS) to establish survey networks allows for the successful integration and comparison of 3D data sets (Figure 9.4). This is particularly important when working in the coastal zone (Figure 9.5) where it is traditionally difficult to identify stable permanent survey reference points to monitor coastal change and where there can be insufficient cellular internet coverage for real-time measurement corrections. To address these challenges both post-processed kinematic (PPK) and real-time kinematic (RTK) methods were used to establish networks. This combination of methods provided a robust and versatile approach to using GNSS at the coast, ensuring each site was recorded to the same level of detail regardless of the method used (RICS 2010).

To ensure that any future, post-CHERISH surveys used to identify change are aligned with data captured during the project, permanent survey markers were also established at several priority sites using GNSS.

FIGURE 9.4 Member of the CHERISH team carrying out a RTK GNSS survey of the eroding coastline of Bardsey Island, Gwynedd.

UPSTANDING HISTORIC STRUCTURES—
TERRESTRIAL LASER SCANNING (TLS)

Terrestrial laser scanning (TLS) was developed towards the end of the 20th century,
largely to conduct as-built surveys of complicated industrial complexes (Ebrahim
2014). However, the potential value in 3D documentation of archaeology and
built heritage was soon apparent. Terrestrial laser scanners are contact-free, non-
destructive measuring devices that record objects in the form of hyper-dense data
sets of individual measurements commonly referred to as point clouds. Each of these
points is assigned x,y,z coordinates, a reflectance/intensity value, and, after processing, an RGB value. Complete structures are captured by combining scans from multiple locations, registered (joined together to form one data set) by using common
targets in individual scans, or, increasingly, as manufacturer's software becomes
more sophisticated, by automatic cloud-to-cloud–based registration.

There are several applications where monitoring heritage sites in the coastal environment by TLS presents significant advantages. Being highly accurate (~2 mm) and if
selected, high-density point clouds (<5 mm spacing) can represent structures in detail
(Figure 9.6), documenting their complex geometries and creating a time-stamped
record. The value of such data is immense: documentation at this resolution highlights
the irregularities associated with most historic structures and can offer the potential
to unravel construction phasing for researchers. Further detailed analysis of the point
cloud can contribute to conservation. Crack lines or slumping may be detected, and
through cloud comparison of repeat surveys, even minor changes can be detected

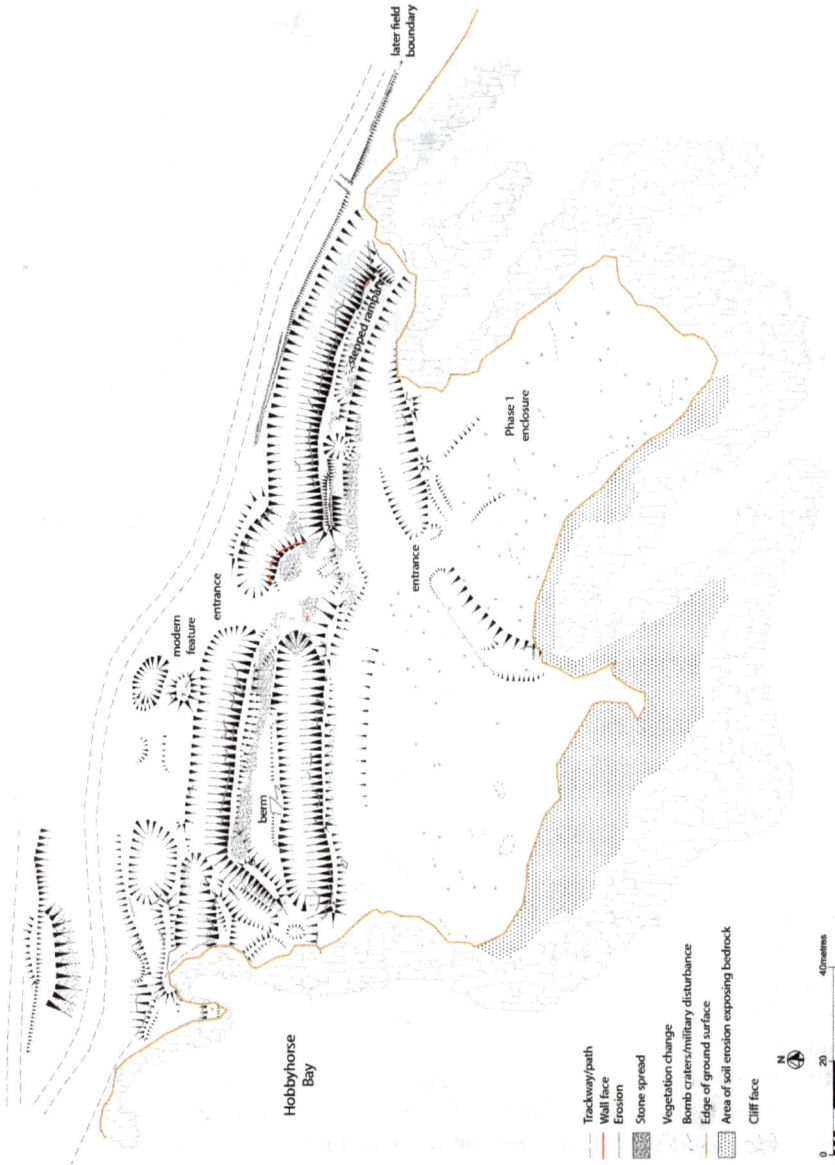

FIGURE 9.5 Archaeological plan generated from GNSS survey of Linney Head Promontory Fort, Pembrokshire.

FIGURE 9.6 TLS-derived 3D point cloud of McCarthy's Castle, Ballinskelligs, Co. Kerry.

(Meneely 2009). Being a remote sensing technique, hazardous, often inaccessible structures can be surveyed from a safe location.

Although the technique is firmly embedded in the toolbox of 3D documentation methods, it does have limitations and present challenges. No 3D laser scan survey is ever 100% complete; data gaps due to laser shadow effects are often significant and can compromise the value of the survey (Grussenmeyer et al. 2008). An example would be the tops of walls or roof detail which cannot be seen from a ground station set up of the scanner. Solutions exist through integration of UAV-generated survey data (Xu et al. 2014), but this requires additional equipment and expertise. Planning the survey to minimise the impact of these gaps is a considerable challenge for the surveyor. Georeferencing 3D point clouds significantly enhances their value, particularly in the context of baseline surveys against which future change can be measured. This can be achieved using GNSS surveying of targets, with fixed permanent markers established as an additional valuable resource (Sairuamyat et al. 2020). Interaction and interrogation of 3D point clouds remains a significant challenge. A level of experience and expertise is required to use 3D viewing software or 3D interfaces, and consequently, users are often more comfortable with conventional outputs such as plans and sections.

As part of the CHERISH project, georeferenced 3D point cloud data sets have been established by TLS for 14 sites of archaeological or cultural heritage significance (Figure 9.7). These data sets are archived with the appropriate metadata to ensure they fulfil the function of a baseline survey against which future change may be detected.

FIGURE 9.7 Operation of phased-based TLS in the field by CHERISH staff recording an eroding cliff face.

3D point clouds may be further processed into 3D textured surface models for re-use in public engagement or educational resources. They provide rich research resources for future understanding and are an important management and conservation resource for future management of the site.

COASTAL RECORDING—MOBILE LASER SCANNING (MLS)

Terrestrial mobile laser scanning (MLS) is achieved by mounting an ALS sensor(s) on a vehicle (Barber et al. 2008), boat (Vaaja et al. 2011), or backpack (Sayama et al. 2019). Techniques have developed over the last two decades (Kukko et al. 2012), and MLS has become widely applied for acquiring high-resolution 3D topographic data in urban (Susaki and Kubota 2017), natural (Vaaja et al. 2011), and coastal environments (Barber and Mills 2007), with systems capable of measuring up to 106 points per second. MLS systems rely on the integration of an inertial measurement unit (IMU) with GNSS to directly obtain georeferenced point clouds that can rapidly acquire large, cost-effective 3D point clouds over extensive areas. Positional accuracy of scan data can range from 14 cm for backpack units (Sayama et al. 2019) to 2–3 cm for vehicle-mounted systems (Barber et al. 2008; Guan et al. 2016), rising to 8 cm for longer-range systems (Di Stefano et al. 2020). Scanning data can be complemented by synchronous camera images.

The overall accuracy of the georeferenced point cloud is generally determined by the accuracy of the navigation GNSS component systems, which work better in open areas with unobstructed views of satellite constellations (Guan et al. 2016). Systems are best deployed over relatively small areas, where it is not cost-effective to deploy airborne ALS, or complex, corridor environments, where multiple viewpoints are required to record the scene accurately (e.g., beach surface, profile, and lower cliffs). Due to the rapid deployment capabilities, systems operate well in environments where data collection is time-limited (e.g., intertidal/coastal areas) or when a high temporal resolution is required. Due to the nature of laser returns, vegetation classification can be more complex than for airborne systems (Lim and Suter 2009; Susaki and Kubota 2017).

Within the CHERISH project, a Trimble MX2 dual-head MLS system was utilised in North County Dublin on a Polaris Ranger ATV. Beach profiles and eroding sand dunes at several sites have been recorded.

AIRBORNE MAPPING AND MONITORING COASTAL SITES AND LANDSCAPES

UNMANNED AERIAL VEHICLE (UAV) PHOTOGRAMMETRIC SURVEY

Unmanned aerial vehicle (UAV) platforms are a valuable source of image data that can be used in the recording of the historic environment, both in the production of 3D data through post-processing of the imagery and in providing an accessible means of capturing aerial imagery of sites and landscapes. A combination of 'structure from motion' (SfM) photogrammetry for environmental monitoring (Westoby et al. 2012) and recent technological advances have made off-the-shelf UAVs a viable low-cost alternative to airborne survey methodologies (Colomina and Parc Mediterrani de la Tecnologia 2008; Eisenbeiß 2009; Shahbazi et al. 2014), making them more accessible to conservation and environmental researchers and managers (Coveney and Roberts 2017). An image-based UAV survey delivering centimetre-level resolution requires mission planning, ground control points (or checkpoints), image acquisition, camera calibration, image orientations, and post-processing software for 3D information extraction (Remondino et al. 2011). UAVs have been fundamental in the successful capture of detailed 3D data for the CHERISH Project, where they have been used to good effect to navigate the countless challenges encountered in the coastal zone.

Two types of UAV are used by CHERISH: multi-rotor and fixed-wing. Utilising both types has allowed the project to maximise flexibility in the resolution and speed of data capture across many project areas. The rationale for a UAV survey on the project was as follows:

- To capture high-resolution, highly precise, and accurate geolocated 3D baseline data sets for threatened coastal sites and landscapes
- To conduct repeat monitoring surveys over the course of the project
- To capture 3D data for analysis and interpretation of upstanding archaeological remains
- To provide 3D data to be used for public outreach products such as physical and digital models and animations

The applications for both UAVs types for this project are outlined next:

FIGURE 9.8   A selection of some of the fixed-wing and multicopter UAVs employed by the CHERISH project.

Multicopter UAV Surveys

Recent developments in reliable and user-friendly multicopter platforms (Figure 9.8) have led to an exponential rise in their use by the surveying and archaeological communities (Gonçalves and Henriques 2017). Broadly speaking, multicopters are best suited for surveying small-scale sites and landscapes due to their manoeuvrability and capacity to carry powerful sensors. There is, however, a trade-off between these benefits and the amount of time they are able to stay in the air; the heavier the payload, the shorter the airtime (Boon et al. 2017). Increasing the size of the UAV (and, thus, its battery capacity) can, to a degree, redress this balance; however, this makes transport of equipment increasingly difficult, especially when surveying in remote coastal locations.

Multicopter systems were largely employed by the project as a rapid and safe means of surveying archaeological sites situated within a diverse range of coastal landscapes (Figure 9.9). With each type of coastal environment comes different challenges, such as precipitous cliffs, inaccessible caves, coastal stacks, and changing tides. The flexibility of multicopter platforms was ideal for accessing challenging coastal environments, particularly in Pembrokeshire, County Kerry, County Waterford, and County Wexford, where more traditional terrestrial survey methods were unsuitable.

FIGURE 9.9 Mulicopter UAV derived SfM 3D model of Dunbeg Promontory Fort, County Kerry. Left of image displays underlying geometric details, whilst the right of the image displays the photogrammetric orthoimagery texture.

FIXED-WING UAV SURVEY

Fixed-wing UAVs capable of photogrammetric data acquisition can operate in manual, semi-autonomous, and autonomous modes. UAV systems are capable of rapidly delivering high resolution spatial and temporal images for ecological, topographical, geomorphological, vegetation, and erosion applications (Boon et al. 2017; Colomina and Molina 2014; Coveney and Roberts 2017).

Fixed-wing UAVs generally have longer flight endurance (>40 min) and faster flight speeds >15 m/s, resulting in increased ground coverage acquisition per unit of time in air, so they are better applied for surveying large areas (Boon et al. 2017). However, they generally take a lighter payload, have less stable image capturing (resulting in lower data precision), and have increased requirements for takeoff and landing, including a suitable area without obstacles to land without damage (Boon et al. 2017). The outputs from fixed-wing UAV surveys are similar to multi-rotor UAVs, including digital elevation models, orthophotos, contour lines, 3D models, and vector data (Figure 9.10).

AIRBORNE LASER SCANNING (ALS)

Since its adoption as a survey technique during the 1990s, airborne laser scanning (ALS) has transformed into a rapid and cost-effective way of collecting high-resolution 3D data of landscapes (Historic England 2018). ALS survey was identified by the CHERISH Project as an effective way of collecting high-resolution surveys of large coastal landscapes that lacked any pre-existing 3D data sets. The rationale for ALS data capture by CHERISH was threefold:

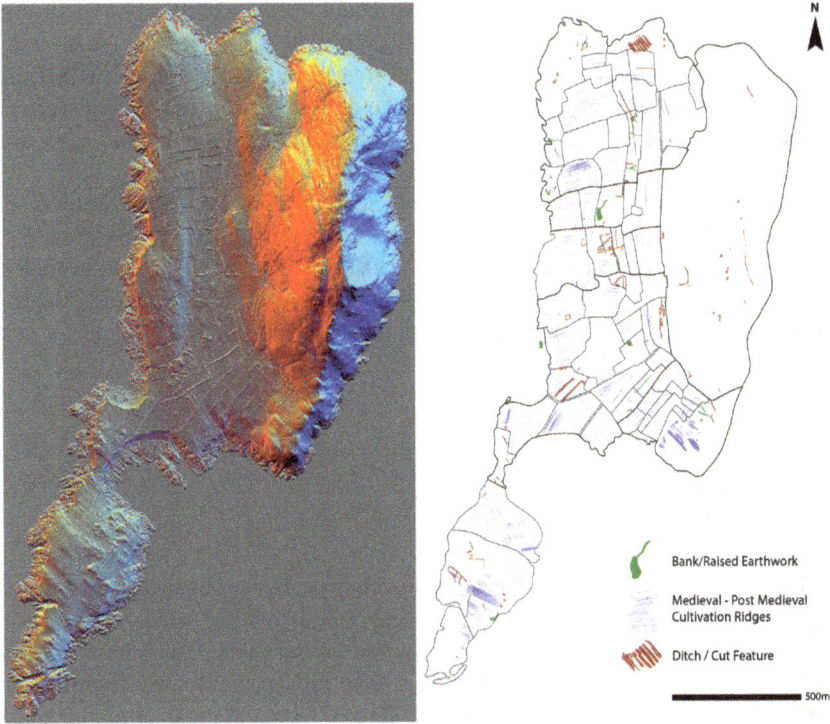

FIGURE 9.10 Fixed-wing UAV survey of Ballinskelligs Bay, Co Kerry, conducted over three days in 2019 covering about 8 km of coastline. Inset of Horse Island at the SW end of the Bay.

- To capture high-resolution 3D baseline data sets for coastal landscapes that lacked any pre-existing 3D data for future coastal erosion monitoring (e.g., six Welsh islands and a section of Dublin's coastline)
- To produce comprehensive archaeological mapping for all upstanding archaeological remains visible within the surveyed landscapes
- To provide data suitable for integration with marine bathymetric data to produce seamless 3D data linking terrestrial and submerged marine landscapes

ALS has been used to good effect by CHERISH in a wide variety of geomorphological and archaeological applications (Driver and Hunt 2018). Its main advantages come with the mapping of human-made, natural features in hillslope, fluvial, glacial, and coastal environments. Its high spatial resolution data over extensive spatial scales (Dong and Qi 2017) allow for subtle topographic features (e.g., archaeological earthworks, river paleochannels, and river terraces) to be identified and mapped more easily, more quickly, and at a lower cost than via traditional field survey methods. Although ground-truthing is still essential (Historic England 2018), ALS also allows for the mapping of remote or inaccessible areas to be undertaken for the production of base maps during desk surveys and for the identification of sites for more targeted, detailed field investigation (Jones et al. 2007). Repeat ALS surveys also allow for temporal change detection (Okyay et al. 2019). For CHERISH, this was fundamental in mapping coastal change and erosion in the dynamic coastal landscapes of both nations, where events such as mass movement and sand dune dynamics were identified.

For each site, an ALS survey was conducted and a series of visualisations that highlight different properties in the data were created based on the raw point data. For this work, visualisations were created using the relief visualisation toolbox (RVT) (Kokalj and Hesse 2017; Kokalj and Somrak 2019). The following visualisations were created for analysis and transcription:

- 16 band multi-directional hillshade
- Slope
- Simple local relief model
- Sky-view factor
- Local dominance

Utilising a range of visualisations aided in providing the best feature interpretation prior to ground-truthing. Each visualisation was imported into a GIS where they could be overlain and manipulated for interpretations and transcription. Using these methods, comprehensive maps of upstanding remains were produced, allowing features to be identified for more in-depth investigation. Figure 9.11 is an example of ALS data and the subsequent transcription of features for Bardsey Island, Gwynedd.

The collection and archiving of 3D ALS data is also important for future monitoring of erosion to coastal heritage. Whilst the resolution of the data (typically between 0.25 m—2 m) does not compare with the much higher resolutions offered by other

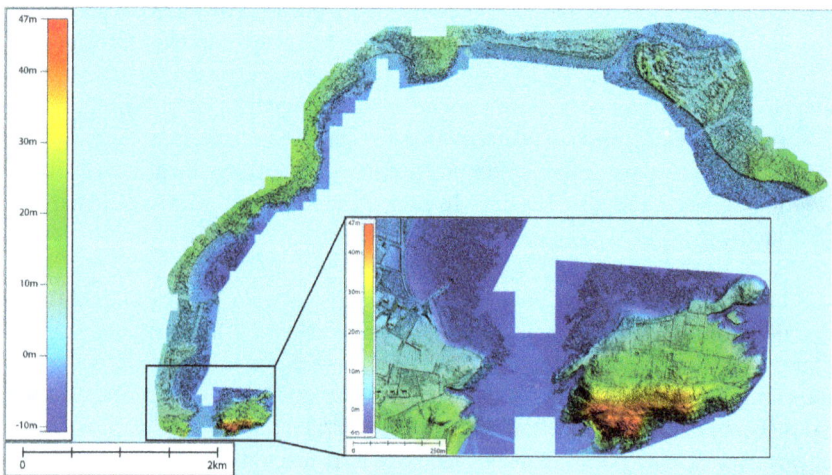

FIGURE 9.11 Multi-Hillshade visualisation of Bardsey Island generated from ALS. Digital transcription of archaeological features visible on the ALS on Bardsey Island, Gwynedd.

methods, such as terrestrial laser scanning or UAV photogrammetry, ALS data sets can be replicated remotely, rapidly, and on a landscape-wide scale. The changes detected through comparison of temporal ALS data sets will only ever be as good as the resolution of the collected data and the accuracy and precision of the GNSS correction data.

MARINE MAPPING AND MONITORING COASTAL SITES AND LANDSCAPES

Marine Geophysics—Multibeam Echosounder Systems (MBES)

Over the last three decades, multibeam echosounder systems (MBES) have become established as the principal means to map large areas of the seafloor for geological, geomorphological, biological, and archaeological purposes (Brown and Blondel 2009; Craven et al. 2021; Kostylev et al. 2001; Quinn and Boland 2010; Westley et al. 2011). Their use has become widespread for coastal and continental shelf investigations due to the simultaneous and continuous collection of both high-resolution bathymetry and backscatter data. This allows site or regional maps to be compiled on which to base more detailed investigations.

Acoustic seabed surveying projects sound energy into the water at a known time and discern the echo from the sediment-water interface. The speed of sound through the water column is used to derive the distance to the reflecting body. MBES transmit multiple acoustic beams (up to >200 depending on system) beneath the vessel in a swath. This allows simultaneous 2D measurements at adjacent beam footprints across the swath width; up to 20 times water depth. As the vessel moves forward, these 2D soundings combine to form a continuous 3D coverage of the seabed. Additionally, measurements of the variation in acoustic backscatter strength provides information on the seabed type (e.g., bedrock gives a stronger backscatter signal than mud).

MBES systems provide cost-effective, continuous acoustic coverage of the seafloor in a range of water depths (from m to km depths), with data coverage and resolution superseding other conventional acoustic survey systems (e.g., side-scan sonar) as a mapping tool (Brown and Blondel 2009). Two factors control the MBES potential bathymetric target resolution capability: distance between soundings and the size of the nadir footprint. Decreasing these results in higher resolution capability (Kenny et al. 2003). However, MBES are relatively expensive compared to other survey techniques (survey-grade systems start about €60k), and due to the large volume of data generated, significant processing of the data is required, with cleaning by experienced personnel to reduce noise (O'Toole et al. 2020). Sonar performance is limited during data acquisition by weather conditions, vessel speed, beam setup, ping period, and sequence, while sources of depth error include sensor configuration, calibration, sound velocity measurements, and tidal corrections that must be identified and minimised (Hughes Clarke et al. 1996). Surveying in shallow water has increased limitations to deeper water due to the increased risk of navigational hazards, decreased swath width, and increased density of depth soundings. This results in the difficult to map 'white

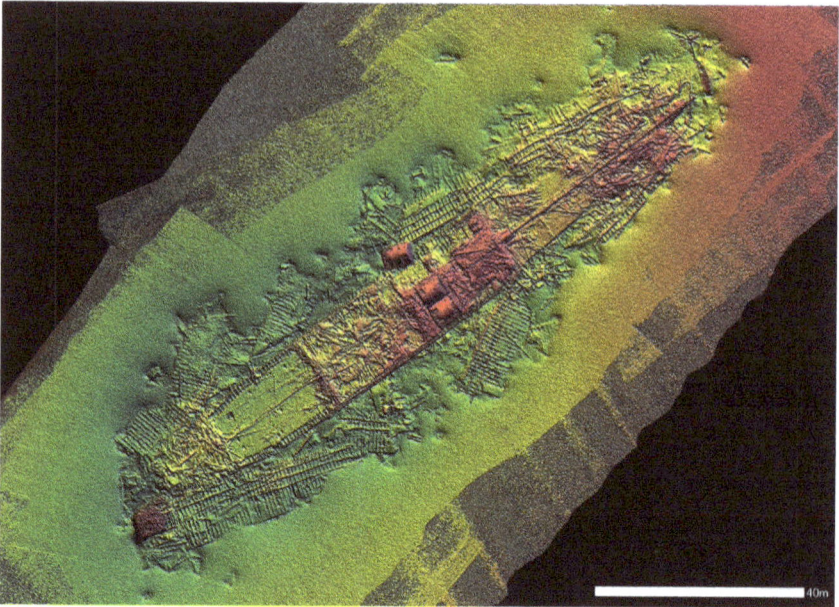

FIGURE 9.12   Multibeam survey of the Manchester Merchant wreck site, Co Kerry, from 2019.

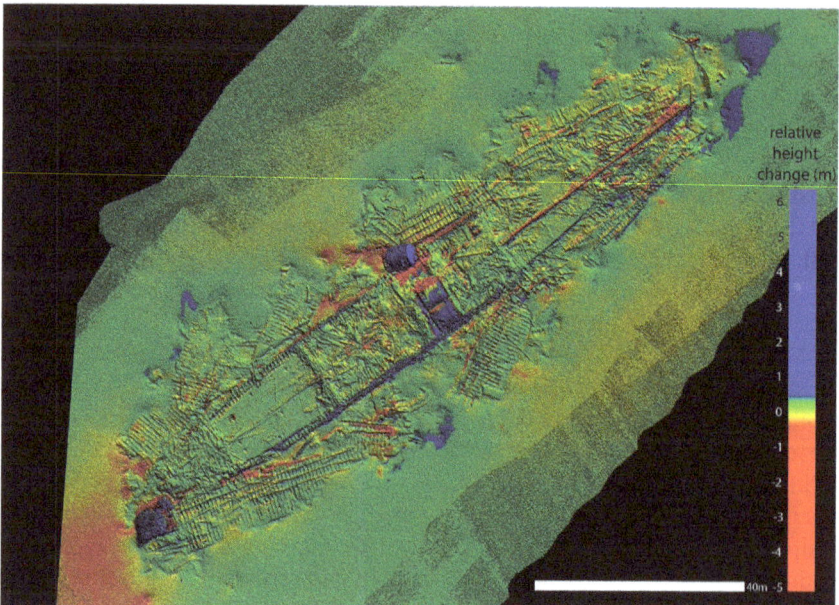

FIGURE 9.13   Analysis of change between the two MBES surveys carried out on the Manchester Merchant wreck site, between 2009 and 2019: erosion in the southwest end of wreck, deposition in the northeast, with degradation of the superstructure including boilers; (prevailing weather is from the southwest. Elevation change scale is from −5.2 m (red) to +6.3 m (blue)).

ribbon' surrounding coastlines, requiring multiple survey methodologies to map (See Seamless Data section).

MBES bathymetric surveying has been deployed by CHERISH in both Ireland and Wales to establish baseline conditions and assess change (Figures 9.12 and 9.13). Target areas have focussed on unmapped shallow marine environments adjacent to sites of cultural significance. Baseline data from some of these sites have been merged with ALS and photogrammetric data to produce seamless coastal maps. Maps of slope, roughness, and relative bathymetry have been produced to provide further contextual information on the seafloor with backscatter data being used to inform submerged sediment type and distribution. Repeat surveys at selected ship-wreck sites are assessing ongoing change.

Remotely Operated Vehicle (ROV) and Diver-Derived Marine Photogrammetry

The introduction of underwater photogrammetric mapping to the marine environment occurred in the early 1960s, developed by leading practitioners including George Bass and Ole Jacobi at the Institute of Photogrammetry and Topography of Karlsruhe University (Whittlesey 1974; Bass 1966). Similarly to UAV survey, the use of SfM in marine archaeological recording has been accelerated by continual camera development and improvement in image quality, positioning systems, and processing software (Remondino et al. 2011; Nex and Remondino 2014). These advances have enabled multi-image photogrammetry of submerged cultural heritage to become commonplace in underwater archaeological investigations, utilising both divers and ROVs for image capture (Waldus et al. 2019; Nornes et al. 2015; Beltrame and Costa 2018; McCarthy and Benjamin 2014; Henderson et al. 2013; Demesticha et al. 2014). To provide local 3D SfM models within a geographical coordinate system, recent studies (Kan et al. 2018) have investigated the use of incorporating precise control points from multibeam echosounder data sets. The CHERISH project captures SfM data through both diver and ROV platforms, with resultant models overlain onto the MBES data set, producing high-resolution and accurate visualisations of underwater cultural heritage sites, which are largely inaccessible to the public.

Marine SfM enables larger areas to be mapped at a high resolution and accuracy in a much shorter timespan than traditional underwater archaeological recording and surveying techniques. Photogrammetric surveys can complement other remote survey such as MBES. Underwater SfM also informs the current condition of the wreck site, its wider environment, and site formation processes while also acting as a tool to monitor sites and rates of change over time. This data capture provides an extensive resource; with the adaption of a more analytical and critical approach to 3D data, it goes far beyond simple measurement and the generation of 2D plans, profiles, and cross-sections. It informs not just individual structures but also their context and environs (Campana 2017). The products from marine photogrammetric surveys include orthoimages, digital surface models, 3D visualisations, and the extraction of metric information. Creating accurate maps and visualising underwater sites is important for the future preservation, long-term study, and use of underwater archaeological sites (Kan et al. 2018).

SATELLITE-DERIVED BATHYMETRY

Marine acoustic surveys using surface vessels enable high-resolution bathymetric data acquisition (O'Toole et al. 2020; Westley et al. 2011). However, these survey methods are costly, time-consuming, and restricted by coastal morphology, navigation hazards, or protected areas. While airborne ALS surveys provide spatial continuity, hardware and operational costs, coupled with logistical requirements, can reduce the frequency of repeat survey. As such, earth observation (EO) has been increasingly used as an alternative to traditional bathymetric survey techniques (Brando et al. 2009; Monteys et al. 2015).

Satellite-derived bathymetry (SDB) from optical satellite images relies on deriving depth data from shallow water environments where light penetrates the water column and from the rate of spectral light attenuation through water. SDB can be divided into two main approaches: empirical and physics-based model inversion. Empirical approaches rely on known bathymetry data points to estimate unknown depths through statistical regression of light attenuation (Lyons et al. 2011; Stumpf et al. 2003), while physics-based methods more tightly constrain unknown depths and attempt to derive them at each pixel in the image, based on water column and bottom substrate reflectance (Dekker et al. 2011; Hedley et al. 2016).

Due to the systematic collection of satellite images over the past decades, extensive image databases exist that can be analysed to provide temporal and spatial continuity, facilitating large areas to be assessed and time series changes to be identified (Hedley 2018).

However, seabed type and conditions of both the water column and atmosphere impact SDB, with this method operating best in calm, shallow, sandy, low turbidity environments with an absence of cloud cover (Hedley 2018; Lyons et al. 2011). Due to fluctuating coastal variables, case-specific modelling techniques for coastlines are required. Therefore, while bathymetry can be modelled over large areas, validation of data is an essential requirement to ensure the accuracy of results.

For CHERISH, suitable optical images were selected from sites around Ireland. Sentinel-2 images were downloaded from the Copernicus Scientific Data Hub website as Level-1C, top-of-atmosphere (TOA) reflectance in 100 km x 100 km tile format. Three 10 m spatial resolution bands were considered (B2–B4) and images were corrected for atmospheric and sunlight effect. Satellite-derived bathymetry was derived using a model inversion method (Casal et al. 2020).

DATA INTEGRATION AND ANALYSIS

SEAMLESS DATA

Coastal sites occur at the transition between terrestrial and marine environments and are affected by processes that occur in both these environments (e.g., fluvial, waves, tidal). Understanding environmental change occurring at coastal sites, therefore, requires accurate integration of both terrestrial and submerged elevation data sets. However, while a key process for mapping, modelling, and forecasting the climate-driven changes to geomorphic processes and environmental responses, these integrated products are difficult to produce due to surveying challenges (and costs) in the shallow water zone (Jiang et al. 2004; Leon et al. 2013; Li et al. 2001; Prampolini et al. 2020).

The 'white ribbon' refers to the dynamic coastal nearshore area at the transition between terrestrial and marine environments, where accurate elevation data is challenging to acquire. A combination of satellite, UAV, terrestrial laser scanning, and acoustic remote sensing technologies can provide data for seamless coastal terrain models (CTMs). However, technical issues, including differences in resolution, precision, and accuracy, can make data integration difficult.

In the CHERISH project, CTMs were produced using data acquired with multiple remote sensing methods deployed through the project. Marine surveying at key coastal sites coincided with high tides to extend bathymetry into shallow water, while UAV photogrammetry surveying and/or aerial ALS data was conducted at times of low water. Data was gridded to a common resolution and vertical elevation datum and merged using a geographic information systems (GIS) software package to produce seamless onshore/offshore CTMs (Figure 9.14).

At other localities, satellite-derived bathymetry, extending to 8 m water depth, was generated from suitable satellite images (cloud and turbidity free at times of high water) and validated using up-to-date bathymetry data. This satellite-derived

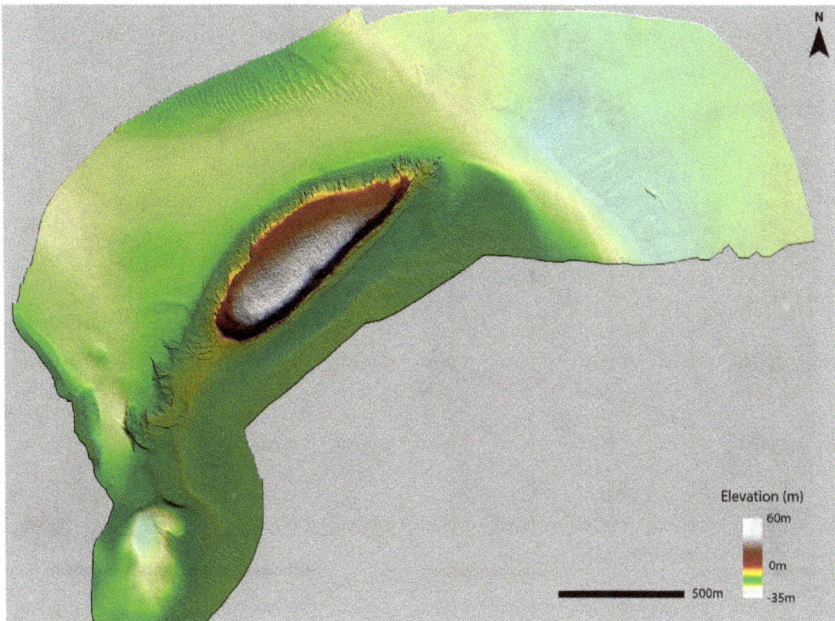

FIGURE 9.14 Seamless integration of ALS- and MBES-derived elevation models from Puffin Island, North Wales.

bathymetry was merged with terrestrial topographic and marine acoustic bathymetric data to create seamless maps.

CHANGE DETECTION

Coastal sites are under constant geomorphological evolution, with seasonal and tidal changes continually altering these habitats. Climate change is expected to increase

FIGURE 9.15 'DEM of Difference' (DoD) analysis for UAV-derived DEM between 2018 and 2019 at Glascarrig Motte and Bailey site on the County Wexford coastline.

the number and frequency of droughts, storms, and heavy precipitation along with rising sea levels, exacerbating this change (IPCC 2018). Historical records, oral testimonies, and early to modern mapping have provided real evidence of change in recent history (Pollard et al. 2020). Therefore, to study the impact of climate change on the coast, remote sensing techniques are required to establish baseline data from which to measure this change (Micallef et al. 2013; Tysiac 2020). Recording and understanding geomorphic evolution is also important for coastal managers and planners, particularly with regard to climate-driven future changes (Esposito et al. 2017; Tysiac 2020).

The comparison of sequential DEMs to produce a 'DEM of Difference' (DoD) is a particularly powerful technique (James et al. 2012). This is a 2D analysis that measures vertical offsets in raster images on a pixel-by-pixel basis (Figure 9.15), with change detection constrained by the resolution of the DEM. However, complex topography, often found in eroding coastal sites with vertical slopes, landscape roughness, and overhangs, are not accurately represented in DEMs. Such sites can experience complex change at varying scales and are, therefore, problematic for change analysis using the DoD method. In these cases, direct comparison between 3D point clouds to evaluate change can be advantageous, although uncertainties can still remain (Esposito et al. 2017; Lague et al. 2013).

The CHERISH project has applied both these methodologies to study change on its sites. DoDs were produced using GIS software for local landscape change analysis (up to 2 km), where impacts of landscape roughness and gridded cell resolution are reduced. At higher scales on individual sites (up to 500 m), volumetric assessments of change on eroding cliff faces were compared using cloud-to-cloud algorithms.

UAV photogrammetry surveys of promontory forts in Co. Waterford and Co. Kerry have revealed missing sections of embankment defences and hut sites with the walls running over the cliff. In the case of the latter, cliff collapse has been recorded over the course of the project. In Waterford, these hut sites on sea stacks and islets would originally have been connected to the mainland around 1,500 years ago when the forts were inhabited (Pollard et al. 2020).

REFERENCES

Barber, D. M., and Mills, J. P. 2007. Vehicle based waveform laser scanning in a coastal environment. *International Archives of Photogrammetry, Remote Sensing and Spatial Information Sciences* 36, no. part 5: C55.

Barber, D. M., Mills, J. P., and Smith-Voysey, S. 2008. Geometric validation of a ground-based mobile laser scanning system. *ISPRS Journal of Photogrammetry and Remote Sensing* 63: 128–141. https://doi.org/10.1016/j.isprsjprs.2007.07.005

Bass, G. F. 1966. *Archaeology Under Water.* London: Thames & Hudson.

Beltrame, C., and Costa, E. 2018. 3D survey and modelling of shipwrecks in different underwater environments. *Journal of Cultural Heritage* 29: 82–88.

Boehler, W., Heinz, G., and Marbs, A. 2001. The potential of noncontact close range laser scanners for cultural heritage recording. *Proceedings XVIII ICIPA Symposium, Potsdam, Germany.* http://cipa.icomos.org/fileadmin/papers/potsdam/2001-11-wb01.pdf

Boon, M. A., Drijfhout, A. P., and Tesfamichael, S. 2017. Comparison of a fixed-wing and multi-rotor UAV for environmental mapping applications: A case study. *The International Archives of the Photogrammetry, Remote Sensing and Spatial Information Sciences* XLII-2/W6: 47–54.

Brando, V. E., Anstee, J. M., Wettle, M., et al. 2009. A physics based retrieval and quality assessment of bathymetry from suboptimal hyperspectral data. *Remote Sensing of Environment* 113: 755–770. https://doi.org/10.1016/j.rse.2008.12.003

Brown, C. J., and Blondel, P. 2009. Developments in the application of multibeam sonar backscatter for seafloor habitat mapping. *Applied Acoustics* 70: 1242–1247. https://doi.org/10.1016/j.apacoust.2008.08.004

Campana, S. 2017. Drones in archaeology. State-of-the-art and future perspectives. *Archaeological Prospection* 24: 275–296. https://doi.org/10.1002/arp.1569

Casal, G. H., Hedley, J. D., Monteys, X., et al. 2020. Satellite-derived bathymetry in optically complex waters using a model inversion approach and Sentinel-2 data. *Estuarine, Coastal and Shelf Science* 241: 106814. https://doi.org/10.1016/j.ecss.2020.106814

Colomina, I., and Molina, P. 2014. Unmanned aerial systems for photogrammetry and remote sensing: A review. *ISPRS Journal of Photogrammetry and Remote Sensing* 92: 79–92. https://doi.org/10.1016/j.isprsjprs.2014.02.013

Colomina, I., and Parc Mediterrani de la Tecnologia. 2008. Towards a new paradigm for high-resolution low-cost photogrammetry and remote sensing. *The International Archives of the Photogrammetry, Remote Sensing and Spatial Information Sciences, ISPRS Congress, Beijing, China, XXXVII. Part B* 1: 1201–1206. www.isprs.org/proceedings/xxxvii/congress/1_pdf/205.pdf

Coveney, S., and Roberts, K. 2017. Lightweight UAV digital elevation models and orthoimagery for environmental applications: data accuracy evaluation and potential for river flood risk modelling. *International Journal of Remote Sensing* 38: 3159–3180. https://doi.org/10.1080/01431161.2017.1292074

Craven, K. F., McCarron, S., Monteys, X., and Dove, D. 2021. Interaction of multiple ice streams on the Malin Shelf during deglaciation of the last British—Irish ice sheet. *Journal of Quaternary Science* 36: 153–168. https://doi.org/10.1002/jqs.3266

Dekker, A. G., Phinn, S. R., Anstee, J., et al. 2011. Intercomparison of shallow water bathymetry, hydro-optics, and benthos mapping techniques in Australian and Caribbean coastal environments: Intercomparison of shallow water mapping methods. *Limnology and Oceanography Methods* 9: 396–425. https://doi.org/10.4319/lom.2011.9.396

Demesticha, S., Skarlatos, D., and Neophytou, A. 2014. The 4th-century B.C. shipwreck at Mazotos, Cyprus: New techniques and methodologies in the 3D mapping of shipwreck excavations. *Journal of Field Archaeology* 39: 134–150. https://doi.org/10.1179/0093469014Z.00000000077

Di Stefano, F., Cabrelles, M., García-Asenjo, L., Lerma, J. L., et al. 2020. Evaluation of long-range mobile mapping system (MMS) and close-range photogrammetry for deformation monitoring. A case study of Cortes de Pallás in Valencia (Spain). *Applied Science* 10, no. 19: 6831. https://doi.org/10.3390/app10196831

Dong, P., and Qi, C. 2017. *LiDAR Remote Sensing and Applications*. USA: CRC Press, Florida, USA.

Driver, T., and Hunt, D. 2018. The white ribbon zone. *RICS Land Journal*, February/March: 22–23. https://issuu.com/ricsmodus/docs/land_journal_february_march_2018/3

Ebrahim, M. A. 2014. 3D laser scanners' techniques overview. *International Journal of Science and Research (IJSR)* 4, no. 10: 323–331.

Eisenbeiß, H. 2009. *UAV Photogrammetry*. PhD diss., ETH Zurich. https://doi.org/10.3929/ETHZ-A-005939264

Esposito, G., Salvini, R., Matano, F., et al. 2017. Multitemporal monitoring of a coastal land-slide through SfM-derived point cloud comparison. *The Photogrammetric Record* 32: 459–479. https://doi.org/10.1111/phor.12218

Gonçalves, J. A., and Henriques, R. 2017. UAV photogrammetry for topographic monitoring of coastal areas. *ISPRS Journal of Photogrammetry and Remote Sensing* 104: 101–111. https://doi.org/10.1016/j.isprsjprs.2015.02.009

Grussenmeyer, P., Tania, L., Voegtle, T., et al. 2008. Comparison methods of terrestrial laser scanning, photogrammetry and tacheometry data for recording of cultural heritage buildings. *International Archives of Photogrammetry, Remote Sensing and Spatial Information Sciences* 37/B5: 213–218.

Guan, H., Li, J., Yu, Y., and Liu, Y. 2016. Geometric validation of a mobile laser scanning system for urban applications. *Proc. SPIE 9901, 2nd ISPRS International Conference on Computer Vision in Remote Sensing (CVRS 2015)*, 990108. https://doi.org/10.1117/12.2234907

Guisado-Pintado, E., Jackson, D. W., and Rogers, D. 2019. 3D mapping efficacy of a drone and terrestrial laser scanner over a temperate beach-dune zone. *Geomorphology* 328: 157–172.

Hedley, J. D., Roelfsema, C., Brando, V., et al. 2018. Coral reef applications of Sentinel-2: Coverage, characteristics, bathymetry and benthic mapping with comparison to Landsat 8. *Remote Sensing of Environment* 216: 598–614. https://doi.org/10.1016/j.rse.2018.07.014

Hedley, J. D., Russell, B., Randolph, K., et al. 2016. A physics-based method for the remote sensing of seagrasses. *Remote Sensing of Environment* 174: 134–147. https://doi.org/10.1016/j.rse.2015.12.001

Henderson, J., Pizarro, O., Johnson-Roberson, M., et al. 2013. Mapping submerged archaeological sites using stereo-vision photogrammetry. *International Journal of Nautical Archaeology* 42: 243–256. https://doi.org/10.1111/1095-9270.12016

Historic England. 2018. *3D Laser Scanning for Heritage: Advice and Guidance on the Use of Laser Scanning in Archaeology and Architecture*. Swindon, Historic England. https://historicengland.org.uk/advice/technical-advice/recording-heritage/

Hughes Clarke, J. E., Mayer, L. A., and Wells, D. E. 1996. Shallow-water imaging multibeam sonars: A new tool for investigating seafloor processes in the coastal zone and on the continental shelf. *Marine Geophysical Researches* 18: 607–629. https://doi.org/10.1007/BF00313877

IPCC, 2018. Global warming of 1.5°C. In V. Masson-Delmotte, P. Zhai, H. O. Pörtner, D. Roberts, J. Skea, P.R. Shukla, A. Pirani, W. Moufouma-Okia, C. Péan, R. Pidcock, S. Connors, J. B. R. Matthews, Y. Chen, X. Zhou, M. I. Gomis, E. Lonnoy, T. Maycock, M. Tignor, and T. Waterfield (eds.), *An IPCC Special Report on the Impacts of Global Warming of 1.5°C above Pre-Industrial Levels and Related Global Greenhouse Gas Emission Pathways, in the Context of Strengthening the Global Response to the Threat of Climate Change, Sustainable Development, and Efforts to Eradicate Poverty*. In Press.

James, L. A., Hodgson, M. E., Ghoshal, S., et al. 2012. Geomorphic change detection using historic maps and DEM differencing: The temporal dimension of geospatial analysis. *Geomorphology* 137: 181–198. https://doi.org/10.1016/j.geomorph.2010.10.039

Jiang, Y. W., Wai, O. W. H., Hong, H. S., and Li, Yok Sheung. 2004. A geographical information system for marine management and its application to Xiamen Bay, China. *Journal of Coastal Research*: 254–264.

Jones, A. F., Brewer, P. A., Johnstone, E., and Macklin, M. G. 2007. High-resolution interpretative geomorphological mapping of river valley environments using airborne LiDAR data. *Earth Surface Processes and Landforms* 32: 1574–1592. https://doi.org/10.1002/esp.1505

Kan, H., Katagiri, C., Nakanishi, Y., et al. 2018. Assessment and significance of a World War II battle site: Recording the USS Emmons using a high-resolution DEM combining multibeam bathymetry and SfM photogrammetry. *International Journal of Nautical Archaeology* 47: 267–280. https://doi.org/10.1111/1095-9270.12301

Kenny, A. J., Cato, I., Desprez, M., et al. 2003. An overview of seabed-mapping technologies in the context of marine habitat classification. *ICES Journal of Marine Science* 60, no. 2: 411–418. https://doi.org/10.1016/S1054-3139(03)00006-7

Kokalj, Ž., and Hesse, R. 2017. Airborne laser scanning raster data visualization: A guide to good practice. *Založba ZRC* 14.

Kokalj, Ž., and Somrak, M. 2019. Why not a single image? Combining visualizations to facilitate fieldwork and on-screen mapping. *Remote Sensing* 11, no. 7: 747. https://doi:10.3390/rs11070747

Kostylev, V. E., Todd, B. J., Fader, G. B. J., et al. 2001. Benthic habitat mapping on the Scotian Shelf based on multibeam bathymetry, surficial geology and sea floor photographs. *Marine Ecology Progress Series* 219: 121–137.

Kotilainen, A. T., and Kaskela, A. M. 2017. Comparison of airborne ALS and shipboard acoustic data in complex shallow water environments: Filling in the white ribbon zone. *Marine Geology* 385: 250–259.

Kukko, A., Kaartinen, H., Hyyppä, J., et al. 2012. Multiplatform mobile laser scanning: Usability and performance. *Sensors* 12, no. 9: 11712–11733. https://doi.org/10.3390/s120911712

Lague, D., Brodu, N., and Leroux, J. 2013. Accurate 3D comparison of complex topography with terrestrial laser scanner: Application to the Rangitikei canyon (N-Z). *ISPRS Journal of Photogrammetry and Remote Sensing* 82: 10–26. https://doi.org/10.1016/j.isprsjprs.2013.04.009

Leon, J. X., Phinn, S. R., Hamylton, S., et al. 2013. Filling the 'white ribbon'—a multi-source seamless digital elevation model for Lizard Island, northern Great Barrier Reef. *International Journal of Remote Sensing* 34, no. 18: 6337–6354. http://doi.org/10.1080/01431161.2013.800659

Li, R., Liu, J.-K., and Yaron, F. 2001. Spatial modeling and analysis for shoreline change detection and coastal erosion monitoring. *Marine Geodesy* 24: 1–12. https://doi.org/10.1080/01490410121502

Lim, E. H., and Suter, D. 2009. 3D terrestrial ALS classifications with super-voxels and multi-scale Conditional Random Fields. *Computer Aided Design* 41, no. 10: 701–710.

Lyons, M., Phinn, S., and Roelfsema, C. 2011. Integrating quickbird multi-spectral satellite and field data: Mapping bathymetry, seagrass cover, seagrass species and change in Moreton Bay, Australia in 2004 and 2007. *Remote Sensing* 3: 42–64. https://doi.org/10.3390/rs3010042

McCarthy, J., and Benjamin, J. 2014. Multi-image photogrammetry for underwater archaeological site recording: An accessible, diver-based approach. *Journal of Maritime Archaeology* 9: 95–114. https://doi.org/10.1007/s11457-014-9127-7

Meneely, J. 2009. Mapping, monitoring and visualising built heritage, A future for Northern Ireland's built heritage. *Environmental Fact Sheet* 7, no. 4. www.nienvironmentlink.org/cmsfiles/files/Publications/A-Future-for-Northern-Irelands-Built-Heritage.pdf (accessed 9 Jan. 2021).

Micallef, A., Foglini, F., Le Bas, T., et al. 2013. The submerged paleolandscape of the Maltese Islands: Morphology, evolution and relation to Quaternary environmental change. *Marine Geology* 335: 129–147. https://doi.org/10.1016/j.margeo.2012.10.017

Monteys, X., Harris, P., Caloca, S., and Cahalane, C. 2015. Spatial prediction of coastal bathymetry based on multispectral satellite imagery and multibeam data. *Remote Sensing* 7: 13782–13806. https://doi.org/10.3390/rs71013782

Nex, F., and Remondino, F. 2014. UAV for 3D mapping applications: A review. *Applied Geomatics* 6: 1–15. https://doi.org/10.1007/s12518-013-0120-x

Nornes, S. M., Ludvigsen, M., Ødegard, Ø., et al. 2015.Underwater photogrammetric mapping of an intact standing steel wreck with ROV. *IFAC-PapersOnLine* 48, no. 2: 206–211. https://doi.org/10.1016/j.ifacol.2015.06.034

Okyay, U., Telling, J., Glennie, C. L., and Dietrich, W. E. 2019. Airborne lidar change detection: An overview of earth sciences applications. *Earth-Science Reviews* 198: 102929. https://doi.org/10.1016/j.earscirev.2019.102929

O'Toole, R., Judge, M., Sacchetti, F., et al. 2020. Mapping Ireland's coastal, shelf and deep-water environments using illustrative case studies to highlight the impact of seabed mapping on the generation of blue knowledge. *Geological Society, London, Special Publications* 505: 207. https://doi.org/10.1144/SP505-2019-207

Pollard, E., Corns, A., Henry, S., et al. 2020. Coastal erosion and the promontory fort: Appearance and use during late Iron Age and early medieval County Waterford, Ireland. *Sustainability* 12: 5794. http://doi.org/10.3390/su12145794

Prampolini, M., Savini, A., Foglini, F., et al. 2020. Seven good reasons for integrating terrestrial and marine spatial datasets in changing environments. *Water* 12: 2221. https://doi.org/10.3390/w12082221

Quinn, R., and Boland, D. 2010. The role of time-lapse bathymetric surveys in assessing morphological change at shipwreck sites. *Journal of Archaeological Science* 37: 2938–2946. https://doi.org/10.1016/j.jas.2010.07.005

Remondino, F., Barazzetti, L., Nex, F., et al. 2011. UAV photogrammetry for mapping and 3D modeling—current status and future perspectives. *The International Archives of the Photogrammetry, Remote Sensing and Spatial Information Sciences* XXXVIII-1/C22: 25–31. https://doi.org/10.5194/isprsarchives-XXXVIII-1-C22-25-2011

Royal Institution of Chartered Surveyors (RICS). 2010. *Guidelines for the Use of GNSS in Land Surveying and Mapping*, RICS Guidance Note 2nd Edition (GN 11/2010). www.rics.org/globalassets/rics-website/media/upholding-professional-standards/sector-standards/land/guidelines-for-the-use-of-gnss-in-surveying-and-mapping-2nd-edition-rics.pdf

Sairuamyat, P., Peerasit, M., Athisakul, C., et al. 2020. Application of 3D laser scanning technology for preservation and monitoring of Thai pagoda: A case study of Wat Krachee Ayutthaya. *IOP Conference Series: Earth and Environmental Science* 463: 012082. http://doi.org/10.1088/1755-1315/463/1/012082

Sayama, T., Matsumoto, K., Kuwano, Y., et al. 2019. Application of backpack-mounted mobile mapping system and rainfall—runoff—inundation model for flash flood analysis. *Water* 11, no. 5: 963. https://doi.org/10.3390/w11050963

Shahbazi, M., Théau, J., and Ménard, P. 2014. Recent applications of unmanned aerial imagery in natural resource management. *GIScience & Remote Sensing* 51, no. 4: 339–365. https://doi.org/10.1080/15481603.2014.926650

Stumpf, R. P., Holderied, K., and Sinclair, M. 2003. Determination of water depth with high-resolution satellite imagery over variable bottom types. *Limnology and Oceanography* 48: 547–556. https://doi.org/10.4319/lo.2003.48.1_part_2.0547

Susaki, J., and Kubota, S. 2017. Automatic assessment of green space ratio in urban areas from mobile scanning data. *Remote Sensing* 9, no. 3: 215. https://doi.org/10.3390/rs9030215

Tysiac, P. 2020. Bringing bathymetry ALS to coastal zone assessment: A case study in the Southern Baltic. *Remote Sensing* 12: 3740. https://doi.org/10.3390/rs12223740

Vaaja, M., Hyyppä, J., Kukko, A., Kaartinen, H., Hyyppä, H., and Alho, P. 2011. Mapping topography changes and elevation accuracies using a mobile laser scanner. *Remote Sensing* 3, no. 3: 587–600. https://doi.org/10.3390/rs3030587

Waldus, W. B., Verweij, J. F., van der Velde, H. M., et al. 2019. The IJsselcog project: From excavation to 3D reconstruction. *International Journal of Nautical Archaeology* 48, no. 2: 466–494. https://doi.org/10.1111/1095-9270.12373

Westley, K., Quinn, R., Forsythe, W., et al. 2011. Mapping submerged landscapes using multibeam bathymetric data: A case study from the north coast of Ireland. *International Journal of Nautical Archaeology* 40: 99–112. https://doi.org/10.1111/j.1095-9270.2010.00272.x

Westoby, M. J., Brasington, J., and Glasser, N. F., et al. 2012. 'Structure-from-Motion' photogrammetry: A low-cost, effective tool for geoscience applications. *Geomorphology* 179: 300–314. https://doi.org/10.1016/j.geomorph.2012.08.021

Whittlesey, J. H. 1974. Whittlesey foundation field activities. *Journal of Field Archaeology* 1, no. 3–4: 315–322. https://doi.org/10.1179/009346974791491476

Xu, Z., Wu, L., Shen, Y., et al. 2014. Tridimensional reconstruction applied to cultural heritage with the use of camera-equipped UAV and terrestrial laser scanner. *Remote Sensing* 6, no. 11: 10413–10434. https://doi.org/10.3390/rs61110413

10 3D in the Construction of a Full-Scale Replica of St. Patrick's Cross, Downpatrick

Michael King and John Meneely

INTRODUCTION

In certain lighting conditions, glimpses of the intricate stone carving on three granite fragments of a high cross on display to the public at the entrance to Down Cathedral could be seen. These three fragments (Figure 10.1) are made up of two massive adjoining fragments, F1 and F2, forming the horizontal arm and cross-shaft of the crosshead, and have a combined width of 1.4 m, with cross-arm and cross-shaft, both 49 cm wide and 21 cm thick (Figure 10.1, F1 and F2). F1 has signs of step-carved designs on one side and part of a rectangular sunken panel on the other, while F2 has a circular zone of intricate interlace carving on one side and a circular sunken panel on the reverse (Figure 10.2). The third piece, F3, is an almost square fragment decorated with spirals, measuring 51 cm across, 48 cm high, and 18 cm thick, and has the addition of a small tenon (14 cm square and 4 cm deep) in the center of one edge (Figure 10.1, F3). This square fragment resembles the top, vertical arm of a cross, and this section is clearly associated with the large cross-arm, F1, in having a similar rectangular sunken panel on its reverse (Figure 10.2).

In June 2014 the three fragments, which since 1900 had been kept inside Down Cathedral, were lifted and cleaned by professional conservators from Cliveden Conservation to remove debris sticking to them. This process revealed some, but not all, of the heavily weathered details of complex interlace, step patterns, fretwork, and spiral carvings on them.

THE HISTORY OF THE THREE FRAGMENTS

Detailed research was carried out to investigate the history of these three remaining cross fragments to see if they were indeed part of the original cross said to have marked the burial-place of Saint Patrick. This investigation revealed two drawings, showing both sides of a damaged cross, in a sketchbook of Alexander Johns of Carrickfergus, stored in the Public Records Office of Northern Ireland. One is labeled as the 'Remains of St. Patrick's Cross in the Burying Ground of Downpatrick', and the other, of the opposite side of the cross, is dated 6 September

DOI: 10.1201/9780429327575-10

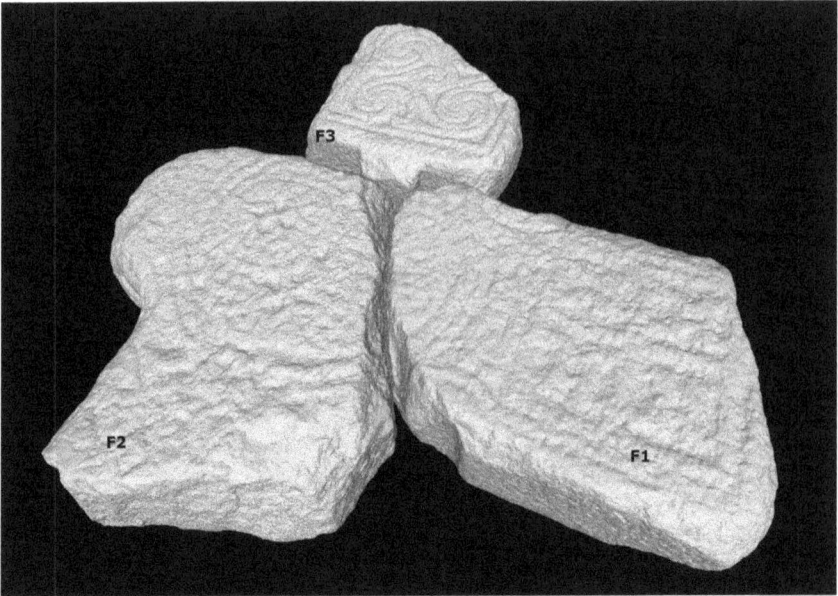

FIGURE 10.1   Front view of the three original cross fragments digitally aligned in Meshlab.

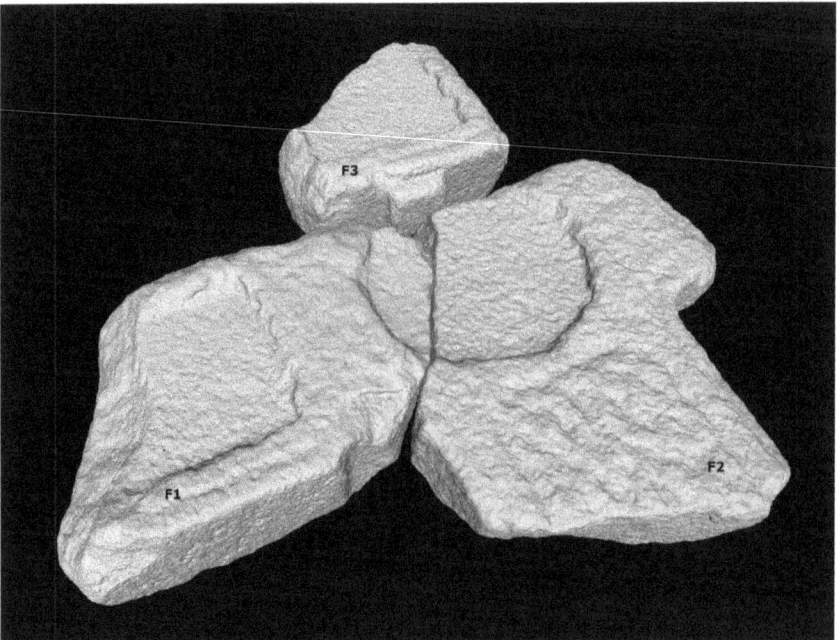

FIGURE 10.2   Rear view of the three original cross fragments digitally aligned in Meshlab.

1843. Two further sketches from an unprovenanced collection of seven pen-and-ink drawings purchased by the Down County Museum, Downpatrick, give possible clues to the detail of the original cross. One of these drawings is captioned 'St Patrick's Cross, Cathedral Burying Grounds', and another uncaptioned sketch illustrates an upright cross-shaft, ornamented with a two-strand twist carved in relief on one side, by the corner of a building. These images post-date several written references to the cross from the 1830s. In 1837, Jonathan Binns of Lancaster describes an earlier visit to the cross in his book 'The Miseries and Beauties of Ireland':

> St. Patrick's Cross, which consists of granite, is a very rude attempt at sculpture, and, in consequence of visitors constantly taking earth from the grave, (for the purpose of keeping it, or selling it in the country for the cure of disease) has a considerable declination from the perpendicular. At the request of the good woman who showed me the relics, I took a piece of the cross for my museum, nor could I disoblige her by refusing to carry away a portion of the soil.
>
> (Binns 1837, 143)

The Ordnance Survey memoirs for County Down include several references to this cross from a visit to Downpatrick by John O'Donovan in 1834:

> In the burial ground connected with the Cathedral a rude fragment of a large granite cross is pointed out as designating the grave of St. Patrick. The portion of the cross which remains is about 3 feet in height and 4 across.
>
> (OS Memoirs 1992, 45)

A further reference in these memoirs refers to a second segment of the cross—

> On an elevated spot in the middle of the graveyard is part of a very ancient cross hewn out of granite, marking the spot where the national saint, St. Patrick, is said to be interred. Another portion of this cross is lying at the corner of the last house on the north side of English Street as you go up to the cathedral.
>
> (OS Memoirs 1992, 45)

This modest property once stood opposite the current route down to the Grove (an area in Downpatrick), and it seems to have been constructed before 1720. This additional segment may have been the cross-shaft at the corner of a property depicted in the uncaptioned image mentioned. In the 23 April 1842 edition of the *Downpatrick Recorder*, it reported—'disgraceful conduct by persons at present unknown, who carried off the headstone from the grave of St. Patrick and conveyed it some distance away on one of the machines employed in making the new road from this town to Ballydugan. The stone was fortunately recovered and stored for safety in the Cathedral'. The cross had obviously been returned to the grave site by September 1843, when Alexander Johns illustrated it, noting substantial damage to one cross-arm. Alexander Knox, in 1875, recorded that—'No sign of the round tower or cross of St. Patrick now remains' (Knox 1875, 417). However, in 1878 O'Laverty remarked—'having been carried off and broken, it was for many years locked up in a portion of the Cathedral', and adds—'the three largest fragments of it are now, however, placed at the east end of the Cathedral' (O'Laverty 1878, 285).

These three remaining fragments appear to have been returned to the supposed grave site sometime before 1891, when a report by Joseph Bigger in November of that year describes a visit by the Secretaries of the Belfast Naturalists' Field Club (BNFC):

> The three fragments at St. Patrick's grave in the Cathedral were first examined and found to compose (almost completely) the head of a heavy massive cross, the remains of that which long stood at the head of the Saint's grave until it was desecrated some years ago. Diligent search was made for the shaft of this Cross but without success. The hope of its ultimate recovery has not been given up for there is a rumor as to its whereabouts in the town sewer where it was thrown. Pending the recovery of the shaft it would scarcely be advisable to restore the cross to its original position.

The fragments of the cross were photographed at the grave site during the 1890s, and at Easter 1900, they were moved back into the Cathedral, prior to the BNFC placing a large granite slab on the site of the supposed grave.

In an article in the Ulster Journal of Archaeology in 1900 (Bigger 1900), it was speculated how the cross may have appeared. An error in this reconstruction is the placing of the approximately square fragment, decorated with spiral carving, at the base of the cross shaft. This fragment is only 18 cm thick, with the other two fragments being 21 cm thick. This square fragment makes more sense as the top arm of the cross, as it could not have supported the two thicker fragments. Also, the small tenon on this square fragment could have fitted into a socket in the top of the damaged T-shaped cross-head (F2) shown in the sketches of the 1840s.

Later research produced differing views on the three remaining cross-fragments, which were, according to the Down and Connor Historical Society Journal (1931), located in the vestry of the Cathedral in 1931. In 1966 The Archaeological Survey of County Down interpreted the largest fragment as the top arm of the cross and identified the spiral-decorated square fragment as part of another cross (ASCD 1966). In 2014, the rediscovery of the drawings from the 1840s supplied crucial evidence for how the three remaining cross fragments could be reconstructed, regardless of the 19th-century damage. One of these drawings shows a spiral design at the top of the cross-shaft supporting the main part of the cross in 1843, suggesting that identical spiral decoration was carved on the shaft and the top arm of the cross. Additionally, the strong possibility that the drawing of the cross-shaft at a corner of a building depicts the fragment of St. Patrick's Cross also described as being at the corner of a building in the Ordnance Survey memoirs makes the design of a 'conjectural replica shaft possible' (King 2020a).

3D LASER SCANNING OF THE CROSS FRAGMENTS

In an attempt to 'reveal' details of the original, now heavily weathered carving on the three remaining fragments of the cross and make this part of the replica as true to the original as possible in detail and scale, the pieces were 3D laser scanned using a tripod-mounted Konica Minolta Vi9i (Figure 10.3). This data was also used to digitally investigate how these remaining fragments may have fitted together.

This high-resolution object scanner captures the 3D geometry of a surface by emitting a horizontal stripe laser beam that sweeps across the surface of an object. Light reflected back from this beam as it travels over and is distorted by any changes in elevation of the surface is received by a charge-coupled device (CCD) sensor, approximately 10 cm above the laser emitter, and then converted by triangulation into distance data. The CCD sensor can also be used with an RGB filter to obtain a colour image of the object once the surface measurement phase is finished. Each scan takes approximately 2.5 seconds and the resulting scanned surface is stored in the form of a dense mesh with mapped texture (if collected). The nominal accuracy of coordinate measurements is 0.05 mm in each direction (Tomaka and Lisniewska-Machorowska 2005).

Numerous, overlapping individual scans were collected on all three cross fragments to ensure that every part of its three-dimensional surface was captured. This was achieved by rotating each piece horizontally on the wheeled plinths on which they were displayed, capturing the top surface and the sides, and then flipping them over and repeating this process. The scanner was not moved during this process. In total 37 scans were collected on F1, 27 on F2, and 23 scans on F3 (Figure 10.1).

These individual scans for each fragment were joined using the vendor-specific software Polygon Editing Tools (PET). This was achieved by selecting three

FIGURE 10.3   The tripod-mounted Konica Minolta Vi9i used in scanning the cross fragments.

common points in two overlapping scans and registering them automatically with the software. Once this successive scan-to-scan registration is complete, a bundle adjustment for all the registered scans on each cross fragment was carried out and a merge function was then used to combine all the individual scans into one single, complete 3D mesh. This merged mesh was checked for any small holes using PET, and, if any existed, they were digitally repaired to make a manifold (watertight) 3D model. Each fragment was exported as a .obj file. The individual 3D models of the cross fragments are available to download for free at https://sketch-fab.com/1manscan/models with a link from each fragment to a 3D model of how they fit together on the same site. These three models of the fragments were then imported into Meshlab (ML), an open-source system for processing and editing 3D triangular meshes.[1] The models were viewed without their colour texture mapped on them—non-photorealistically, as this was found to improve their inspection. One feature in this software and other 3D applications is the ability to direct a light source onto the surface of a 3D model at any angle, in this case by holding down the CTRL and SHIFT keys simultaneously and moving the mouse to change the direction/angle of the light. This relatively simple process can, and did, make some difference in revealing previously unseen detail on the surface. By creating a 3D

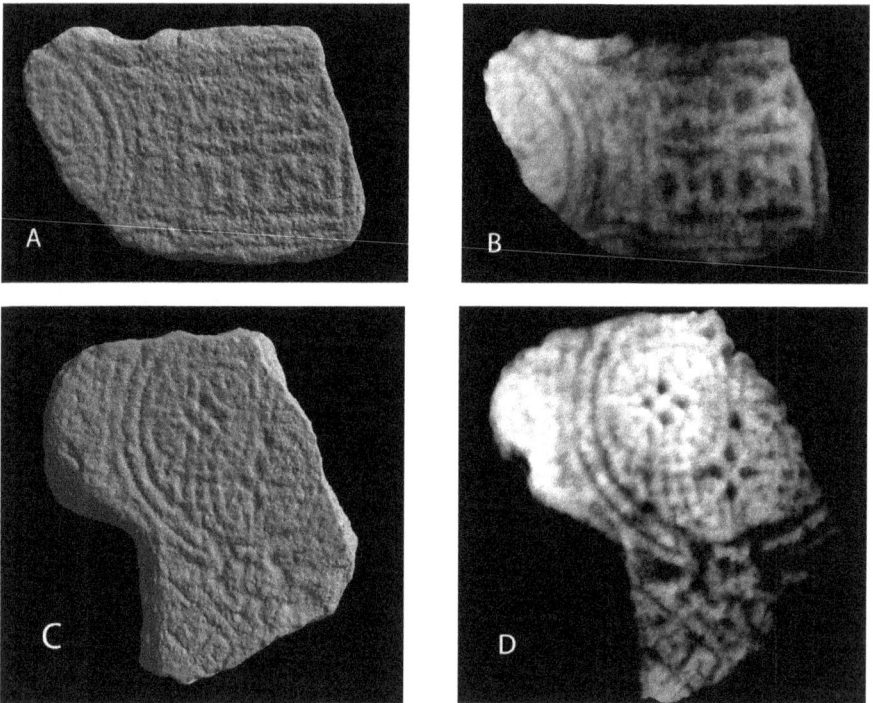

FIGURE 10.4 (A) 3D model of fragment F1 with low-angle light across its surface. (B) Depthmap shader on fragment F1. (C) 3D model of fragment F2 with low-angle light across its surface. (D) F2 Depthmap shader on fragment F2. All images are generated in Meshlab.

model of an eroded carved surface you wish to investigate, especially if the object cannot be moved or traditionally required waiting for the sun to be at a certain angle or additional portable illumination to cast shadows across its surface, this simple technique is extremely useful.

One favoured ML filter for enhancing inscriptions in carved stone is Radiance Scaling (RS)—where reflected lighting variations are correlated to surface feature variations; for example, by increasing reflected light intensity in convex regions and decreasing it in concave ones, the highlight looks as if it were attracted towards convexities and repelled from concavities. Such an adjustment improves the distinction between concave and convex surface features and can aid greatly in improving the visibility of eroded features (Vergne et al. 2010). However, in this case, the RS shader failed to improve on the simple technique of directing light obliquely across the surface to enhance visualising the eroded carving on fragments F2 and F3.

The shader in ML that gave the best results for these stones was Depthmap—a plugin for ML developed by the Visual Computing Lab, Institute of Information Science and Technologies (ISTI), National Research Council of Italy (CNR). This plugin colour codes the z values in a model from white to black, and the range of z values can be easily adjusted with two onscreen sliders zmax and zmin. Adjusting zmin to the lowest point and zmax to the highest point on the surface being investigated greatly improved the visualisation of eroded detail (Figure 10.4). This is particularly evident in F2, where adjusting the direction of light across the surface of this model slightly improved visualising the weather-beaten carving (Figure 10.2 (C)), but this was surpassed by the application of the Depthmap shader, which revealed the complicated interlacing in the centre of the cross arm in much greater detail (Figure 10.2 (D)).

Figure 10.5 shows the difference between using the RS and Depthmap filters on the rear of fragment F2. Due to the very rough surface of these fragments, caused in part by

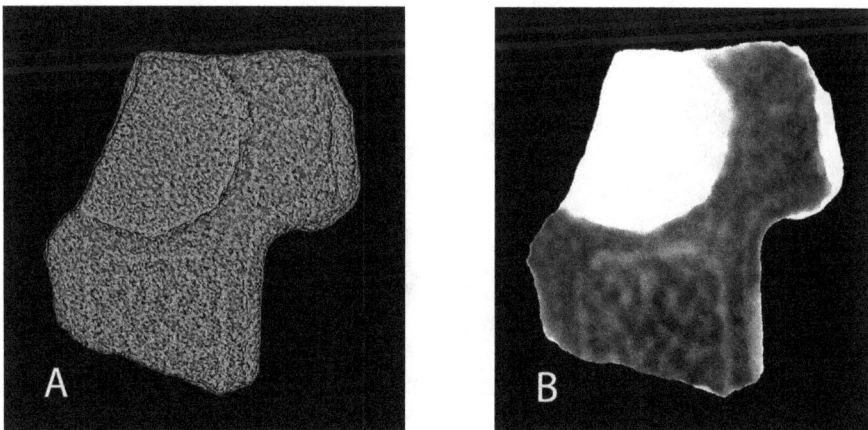

FIGURE 10.5   (A) Radiance Scaling filter applied to the rear of F2. (B) Depthmap filter applied to the rear of F2.

the large crystal size and hardness of the separate minerals of the granite, which erode to produce a very uneven surface, the RS filter illuminated the concave/convex detail at this granular level, impairing the visualisation of the carving, while the Depthmap filter greatly enhanced the carved detail below the circular sunken panel on F2.

Meshlab was also used to digitally manipulate the three fragments to perform a probable best fit and virtually reconstruct these three fragments to show how they may have looked in the past (Figure 10.6). This digital reconstruction of the three fragments combined with the desk-based study gave vital information as to what size the replica should be. Based on these measurements and comparisons of the ratio of span to height with other high crosses, the replica would be approximately 1.6 m wide and 4.7 m tall.

Scaled Depthmap images of the six faces (front and rear) on the three fragments were then generated and sent to a graphic illustrator who carefully created a number of detailed, 1:1 2D scaled drawings for each side of the cross, incorporating the features revealed by the 3D scanning process (Figure 10.7).

The Kilkeel stonemasons S. McConnell & Sons, having previously made a replica of the early 10th-century Mourne granite Town Cross of Downpatrick (King 2014, 2020b), then created a very basic digital model of St Patrick's Cross in four sections—a socket stone, a cross shaft, a T-shaped cross head, and a top arm. These four sections were then cut by a computer numerical control (CNC) machine. The intricately carved designs were finished by hand by a master stone carver, Declan Grant, using the detailed scale drawings generated by the graphic artist (Figure 10.8).

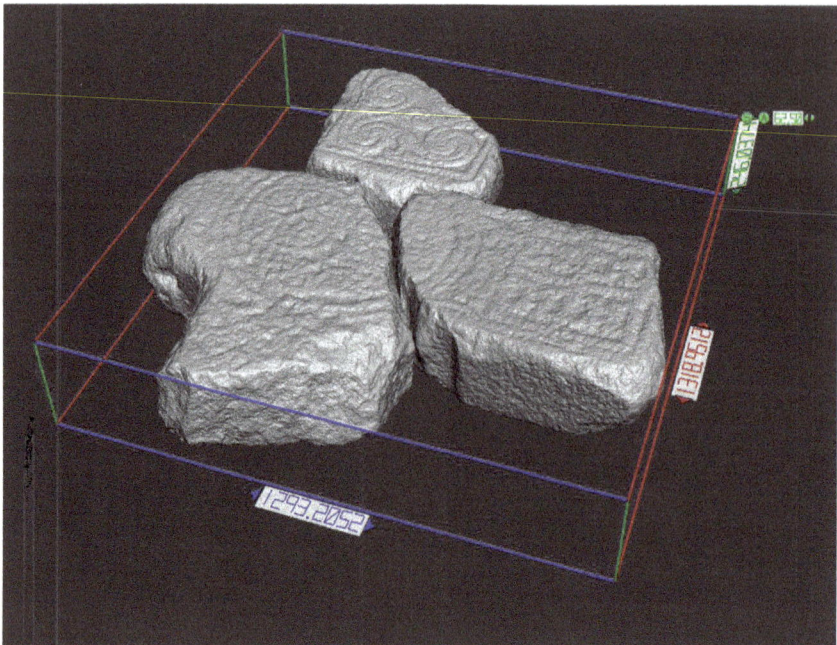

FIGURE 10.6 Dimensions of the digitally reconstructed fragments greatly assisted in deciding the size of the replica.

FIGURE 10.7 One of the scaled drawings produced by the graphic artist Philip Armstrong of what the final carved detail should be on the replica.

FIGURE 10.8 A master stone carver from S. McConnell & Sons hand carving the intricate detail into the cross.

FIGURE 10.9 Construction of the replica St. Patrick's Cross by employees from S. McConnell & Sons.

The replica cross was constructed from the same granite as the original, over 12 centuries previously. These three fragments are carved from G2 granite, which outcrops on the nearby, eastern side of the Mourne Mountains. This is a biotite granite with abundant dark quartz crystals. These were the last granite slabs to be quarried from Thomas's Mountain, which overlooks the town of Newcastle, on the eastern side of the Mourne Mountains in County Down, N. Ireland.

The replica cross, which now stands 4.71 m tall and has a span across the head arms of 1.62 m, was erected directly opposite the entrance to Down Cathedral (Figure 10.9). On Friday, 31 August 2018, the replica cross was formally unveiled, followed by an interdenominational cross-community blessing led by the Bishop of Down.

CONCLUSION

The recreation of St Patrick's Cross was made possible by a combination of funding, new research and professional stone conservation, with graphic representation of the likely form and decoration of the cross made possible by 3D scanning and digital exploration of the original fragments. This made it feasible for master stone carvers to reproduce the cross according to all the fine-detailed information made available to the team by means of the digital modeling methods outlined. It is hoped that the new cross, located opposite the western entrance of Down Cathedral and casting its shadow on October evenings into the doorway each year, where the original fragments lie, will help visitors to re-imagine the Hill of Down in about 800 AD and last a further 1,200 years.

ACKNOWLEDGMENTS

Funding for this research and the production of the replica cross came from Newry, Mourne, and Down District Council, the Friends of Down County Museum, the Lecale and Downe Historical Society, the Dean and Chapter of Down and Fr Fearghal McGrady. Scaled drawings of the carved details for the replica cross were produced by graphic artist Philip Armstrong. The 1840s drawings of the cross were brought to the authors' attention by George Rutherford and Peter Harbison. All work to record the original fragments and install the new cross were by permission of Dean Henry Hull and the Chapter of Down and with the approval of the Historic Environment Division (Department of Communities). The site of the new cross was excavated by the Centre for Archaeological Fieldwork, Queen's University Belfast, under the direction of Brian Sloan. Conventional photographic recording of the fragments was undertaken by Bryan Rutledge. Helpful information was also provided by Megan Henvey, Duane Fitzsimons, Philip Blair and Colm Rooney.

NOTE

1. www.meshlab.net

REFERENCES

An Archaeological Survey of County Down (1966). Archaeological Survey of Northern Ireland.
Bigger, F.J. (1900). The Grave of St Patrick. *Ulster Journal of Archaeology*, vol. 6, pp. 61–64.
Binns, J. (1837). *The Miseries and Beauties of Ireland.*
Down and Connor Historical Society Journal (1931), vol. 4.
King, M. (2014). Moving the Downpatrick High Cross. *Lecale Review*, vol. 12, pp. 7–12.
King, M. (2020a). The Story of St Patrick's Cross. *Lecale Review*, vol. 18, pp. 80–90.
King, M. (2020b). Sacred Granite: Preserving the Downpatrick High Cross. In *Cultures of Stone. An Interdisciplinary Approach to the Materiality of Stone*, eds. G. Cooney, B. Gilhooly, N. Kelly and S. Mallia-Guest, pp. 181–196. Sidestone Press, Leiden, The Netherlands.
Knox, A. (1875). *The History of the County of Down.*

O'Laverty, J. (1878). A Historical Account of the Diocese of Down and Connor. *Ancient and Modern*, vol. 1.

Ordnance Survey Memoirs of Ireland (1992). *Parishes of County Down IV, 1833–37, East Down and Lecale*, vol. 17, eds. Angelique Day and Patrick McWilliams. Institute of Irish Studies, Queen's University, Belfast.

Tomaka, A., and Lisniewska-Machorowska, B. (2005). The Application of the 3D Surface Scanning in the Facial Features Analysis. *Journal of Medical Informatics & Technologies*, vol. 9.

Vergne, R., Pacanowski, R., Barla, P., Granier, X., and Schlick, C. (2010). Radiance Scaling for versatile surface measurement. *Proceedings of the 2010 ACM SIGGRAPH Symposium on Interactive 3D Graphics and Games*, pp. 143–150. Association for Computing Machinery, New York, USA.

11 Thermography Using Unmanned Aerial Vehicles

Scott Harrigan and Harkin Aerial

INTRODUCTION: DEFINITION AND BACKGROUND OF THERMOGRAPHY

Thermal imaging, better known as *thermography* or infrared imaging, is the process of producing an image based on relative amounts of infrared radiation absorbed by a camera sensor. Infrared radiation is a portion of the electromagnetic (EM) spectrum, which includes many other forms of energy such as visible light, radio waves, and X-rays (Figure 11.1). Where typical photographic cameras record and display energy of wavelengths in the visible light portion of this spectrum, thermal cameras do the exact same for energy in wavelengths of the infrared portion.

Thermography has roots further back in history than might be assumed. The first documented study of thermography was performed in 1800 by Sir William Herschel, a German astronomer. Herschel was initially searching for an optical filter to reduce the sun's brightness when using his telescope. Herschel was surprised to discover that some of the materials he had used as optical filters passed wildly varying levels of heat through the telescope to his eye. Driven by the desire to find the perfect filter material—one that dimmed the most sunlight and transmitted the least heat—Herschel started to test the effects of temperature when under the different colors of light split by a prism. Herschel noticed that as he moved a thermometer under each color, the thermometer would rise higher in temperature under the red light when compared to the violet light on the other side of the spectrum. More curiously, as he moved his thermometer further past the red light along the spectrum, into the areas where no visible light was seen, he found a certain distance at which the thermometer peaked (Gromicko and McKenna 2021).

Through further experiments, Herschel made some remarkable and intuitive conclusions that became the bedrock of modern thermal imaging. Firstly, he correctly surmised that radiant heat from the sun was not equally distributed along the energy spectrum. Secondly, he determined there was a certain range of radiant energy wavelengths at which measured heat was the highest. Though it would be centuries until the concepts of energy wavelengths and the electromagnetic spectrum were properly defined, we thermographers owe much to Herschel for inadvertently discovering the infrared wavelengths of energy! (White 2019) Following Herschel, the scientists

DOI: 10.1201/9780429327575-11

THE ELECTROMAGNETIC SPECTRUM

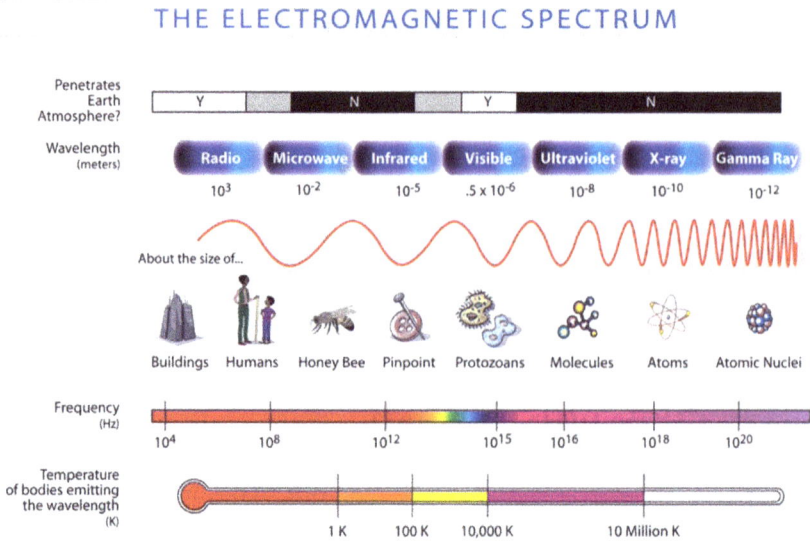

FIGURE 11.1 Diagram of the electromagnetic spectrum. Objects that emit heat in typical temperatures commonly radiate in wavelengths along the infrared portion of the spectrum.

Josef Stefan, Ludwig Boltzmann, Max Planck, and Wilhelm Wien would each make major discoveries about infrared energy that led to our ability to accurately measure and image it today.

THERMAL IMAGING AND UAV TECHNOLOGY TODAY

Advances in camera manufacturing and military applications led to the thermal cameras and unmanned aerial vehicles (UAVs, or drones) of present day. Advancements through the 1940s, 1950s, and 1960s turned thermal cameras from expensive, large, and low-resolution instruments into the precision imaging devices we know them as today. Up to this point, aerial thermography was confined to only large helicopters and airplanes capable of carrying these systems, which required active refrigeration or large detector arrays. Specifically, the invention of the *microbolometer* in 1978 by Honeywell and Texas Instruments was a breakthrough that allowed for low-power, lightweight thermal cameras that did not need active cooling. The microbolometer was a military secret until the early 1990s, when the US military declassified the technology, paving the way for widespread commercial use (Westervelt et al. 2000). The later 1990s and 2000s would see major commercial adoption of these smaller thermal cameras in many industries, such as firefighting, search and rescue, and building envelope inspections.

Around the same period, military UAV technology became further and further miniaturized. The rapid evolution of consumer electronics and the smartphone revolution, combined with this technology, led to the modern consumer and commercial UAVs now prevalent in any electronics store. While the electric motors in today's UAVs borrow directly from the RC hobbyist aircraft that preceded them, many of the components in a modern UAV owe themselves to advances in the low-cost components found in your smartphone. Specifically, lightweight GPS and inertial measurement units (IMUs) common in smartphones provide simple ways for a UAV to determine its location and orientation. High-quality, affordable radio and Wi-Fi transmitters allow for UAV control ranges beyond a mile. Modern smartphone apps also enable sophisticated software to be brought into the field for easy image processing and analysis. Smartphone apps also provide user interfaces tailored to intuitive UAV piloting and aerial photography. Lastly, modern lithium polymer (LiPo) battery technology allows for UAV flight times greater than 30 minutes on a single battery, with considerable payload lifting capability.

With thermal cameras lightweight enough and UAVs powerful enough, it was only a matter of time before hobbyists and professionals started to combine the two. In the mid 2010s, thermal camera manufacturer FLIR and UAV manufacturer DJI partnered to provide ready-to-fly thermal camera/UAV solutions that became the go-to devices for those looking to perform affordable aerial thermography (DJI 2015). Today, thermal cameras are integrated tightly into UAVs small enough to fit in a standard backpack. In the next section, we'll look at how these incredible devices are used regularly today and the many ways one can use thermal technology to benefit one's line of work.

APPLICATIONS AND BENEFITS OF THERMAL UAV TECHNOLOGY

Firefighting may have been the first industry to immediately reap the benefits of thermal UAVs. Since thermal cameras record infrared energy and not visible light, a UAV-mounted thermal camera is able to see directly heat-generating objections through smoke, haze, fog, and most atmospheric obscuration. This is incredibly useful in a house or commercial fire, where smoke obscures roofing and prohibits a bird's-eye view through visible light cameras. For years, institutions such as the Los Angeles Fire Department have successfully implemented thermal drone programs, citing the ability of the technology to "survey hundreds of acres and identify hot spots instead of sending firefighters into harm's way" (Karpowicz 2018).

The **building inspection** industry was an early adopter of handheld thermal camera technology through the 1990s and 2000s. Since then, tech-savvy inspectors have been using thermal cameras to see the heat differences in building walls, floors, and roofs. Heating differences in building envelopes often have two root causes of concern to inspectors—missing insulation or moisture. Missing insulation will cause a portion of a roof or wall to appear hotter than the corresponding insulated sections. By identifying these areas, an inspector can locate insulation to be replaced without requiring attic or roof access. Depending on the time of day, accumulated moisture will appear either hotter or colder than the surrounding dry areas as water has a high specific heat capacity—that is, it takes a wet area longer to heat up or cool down versus a dry area in the same location. Under the right conditions, this difference can be seen clearly, and trained inspectors can identify moisture or leaks early on before they lead to further damage and expensive building repairs. UAVs have greatly enhanced the capabilities of inspectors, as roof thermal imaging can now be performed from the air by a UAV from a handful of photographs instead of many close-up photographs taken by an inspector standing on a roof. This reduces the time of many inspections from hours or days into well under an hour for most roofs. Additionally, walls, HVAC equipment, window seams, and other areas that are difficult to access on foot can be accessed immediately by a UAV with little difficulty. Especially on large area commercial rooftops, such as those in big-box retail stores, UAV technology has proved to be a game-changing technology for professionals tasked with assessing the conditions of these structures.

Environmental monitoring is a rapidly advancing application for UAV thermography. Many issues of environmental concern (such as storm runoff, pollution, drinking water leaks, and natural gas leaks) are strongly associated with a transfer of radiant heat or a mixing of chemicals in water that retain heat at different levels. In most cases, a thermal camera mounted on a UAV can easily detect these heat and chemical variations. Even more so than building inspections, environmental monitoring benefits from the UAV's unique ability to cost-effectively cover large areas in a small amount of time. Miles of coastline and hundreds of acres of land, which once took days or weeks to assess, can be covered by a UAV in as little as a single morning or evening. Recent work in determining illicit discharge detection has proven especially fruitful with a UAV, as small but damaging discharges (such as pollution from basement drains, sump pumps, unmapped outfalls, and other point sources of pollution) can be picked up from high-resolution thermal UAVs flying at low altitudes (Figure 11.2). Not only are manned thermal aircraft relatively expensive, but they are constrained to flying at much higher altitudes. High altitude flights generate less detail, and a manned aircraft may miss all but the largest sources of pollution already known. Accurate image geotagging from the UAV allows users to reference images of issues easily on a map or in GIS software, allowing for quick pinpointing of issues, especially in unfamiliar areas or areas with little visual distinction, such as an open field.

This image was captured in winter in the northeast United States, with an ambient air temperature of approximately 13 °F (-10.56° C). In these conditions, manmade sources of water retain considerably more heat than the surrounding environment and are easy to identify via thermal. Subsequently, heat increases as distance to point of discharge decreases. By following the thermal signature to its relative warmest location, the exact point of discharge can be found. A geo-tagged thermal photo is captured for reference, and a ground follow-up is performed at the recorded location.

The energy industry has long utilized thermal inspections on the ground to identify issues at energy production and distribution facilities. Now with the advent of affordable aerial thermography, components such as flare stacks, oil pipelines, and transmission and distribution lines can be inspected with ease. Previously, these required expensive helicopter-mounted thermal cameras or were inspected less frequently by climbing crews or when components were brought onto the ground as part of other maintenance. Specifically, structures such as distribution pipelines can be inspected to identify defects early and often. By piloting a UAV across a pipeline, defects can be spotted before they become major issues or leaks. This process can be easily repeated, and a UAV can cover miles of pipeline in a single day or night. As the use of UAVs becomes more and more widespread, government aviation authorities have started to issue Beyond Visual Line of Sight (BVLOS) waivers and permissions to a select group of individuals and companies that have a pressing

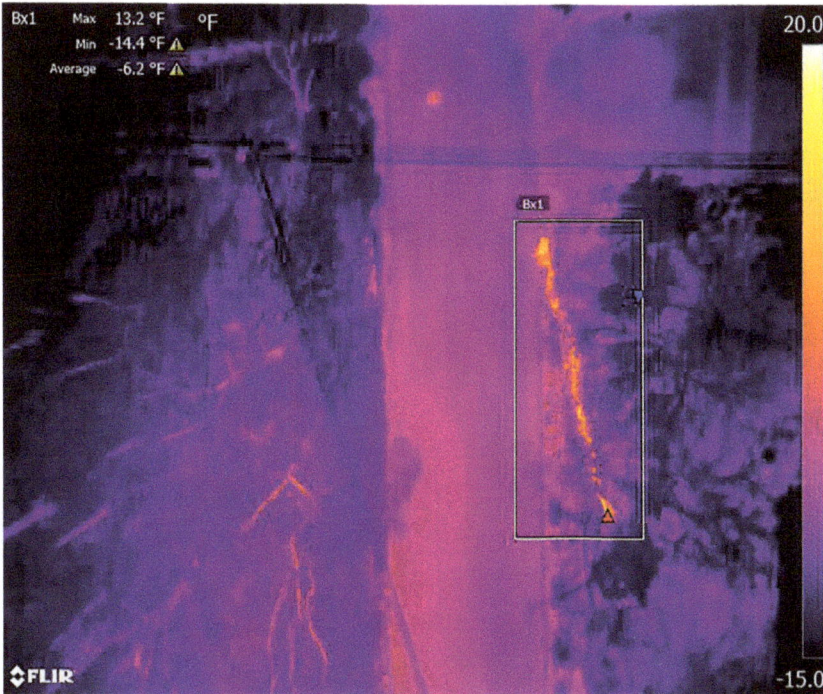

FIGURE 11.2   Thermal image of a residential roadway with a discharge to a storm drain. This image was taken with a UAV in New Paltz, NY, in December 2018.

Source: Reprinted with permission from Harkin Aerial.

need and meet minimum safety standards. A BVLOS waiver grants the holder legal permission to fly a UAV past the point where it can be seen by the piloting crew. Currently, a small handful of energy companies and operators hold BLVOS waivers, but as communication and collision avoidance technologies improve, these types of waivers and permissions could become commonplace, especially for inspections in remote areas.

Green energy technologies have found major benefits from UAV thermography as well. For example, UAV thermal inspections for solar farms are widespread, as a malfunctioning solar cell will generate heat at a different rate than its working counterparts, owing to the increase in resistivity of the component. Thermal imaging can even determine the difference between a single panel failing or an issue in an entire string of multiple solar panels. This technology has proven to be so beneficial that many UAV service providers have started their business solely to serve the rapidly growing needs of solar panel owners and operators.

Lastly, thermal UAV inspections need not even be on a scale as large as solar farms or oil pipelines to prove their value.

From this image (Figure 11.3), it is clear that the transformers are working properly, as heat distribution is even and the gradient is smooth and similar across both

FIGURE 11.3   Thermal image of a properly functioning electrical transformer, taken by a UAV.
Source: Reprinted with permission from Harkin Aerial.

transformers. If an issue were to present itself, it would most likely show up as a single hot spot in the body of the transformer. Additionally, any hot spots in wiring or connectors are an early indicator of corrosion, missing insulation, or an impending short. Rather than operate a single UAV flight across a long distance, in this application, the benefit is being able to simply drive to each pole, quickly unpack the UAV, fly it to the transformer, acquire imagery, land, and repeat. Not only is this method significantly faster than using a ladder or cherry-picker, it also eliminates the possibility of a fall from height (one of the most common and deadly hazards in the construction and inspection industries).

CONSIDERATIONS FOR EFFECTIVE UAV THERMAL IMAGING

Much like photography, thermography can be simple to learn but challenging to master. One of the most important concepts to consider is that thermography is a *passive* measurement, which means it absorbs energy from the environment and subject to make a measurement. It does not emit any light or energy itself. Therefore, acquiring good thermal imagery is entirely dependent on environmental factors at the time of the photo. Such factors include, but are not limited to: *time of day, ambient temperature, temperature of the subject, emissivity of the subject, sun angle, cloud cover,* and *humidity.* In the following section, major considerations for thermal imaging will be discussed, and common techniques to counter these environmental factors will be presented.

Time of day is one of the most wide-ranging factors in thermal imaging. Unlike other factors which may have distinct rules of thumb, the ideal time of day to thermally image will depend completely on the type of application. It is important to first understand if you are trying to see energy being emitted by a system or energy being absorbed by a system, which will dictate if the thermographer desires to find issues during the heating period of the day or the cooling period of the day. For example, if a thermographer's goal is to identify moisture on a flat roof, it is ideal to image at the very end of a sunny day after the sun has heated up the roof. By taking thermal images of the roof right after sunset, the thermographer will properly observe the roof cooling period at different rates: spots with moisture would appear warmer as the water continues to retain heat, while the dryer areas will show up colder as they more effectively bleed off heat in the same timeframe. Were the thermographer to image in the middle of the day, the roof might be in thermal equilibrium and the entire surface might appear to be the same temperature regardless of moisture content.

The opposite would be true for a solar panel inspection, where it would be ideal to image during full operation of the solar panels around mid-day. In this scenario, the thermographer is looking to see solar panels during active hours to see if one is improperly heating up due to issues. In many cases, the goal may be to thermally image a system that has peaks and valleys in equipment operation over time. For example, showers and hand sinks are used most in the morning, so if a domestic water plumbing leak is the thermal subject of interest, the morning may be the best time to image simply because the system in question is at peak operation.

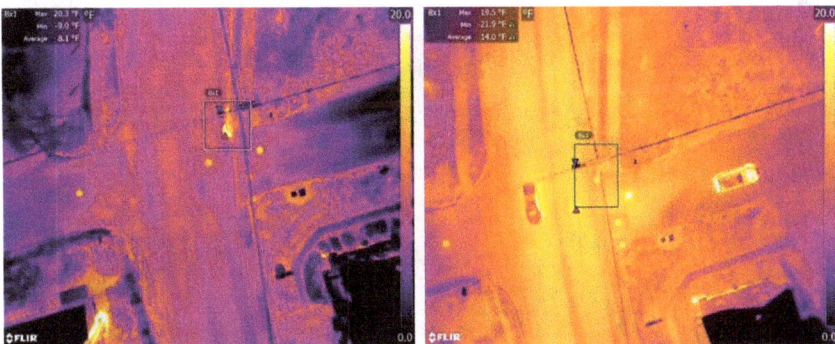

FIGURE 11.4 Side-by-side comparison of two thermal images of the same water main leak. The leak is much easier to spot on the image on the left, which was taken December 18, 2020, at 6:48 a.m., in New Paltz, NY. The corresponding image on the right was taken the same day at 4:36 p.m. In the afternoon, solar loading contributes to a smaller thermal contrast between the leak and roadway. The leak also appears smaller due to lower water demand in the afternoon.

Source: Reprinted with permission from Harkin Aerial.

Figure 11.4 shows a leak is clearly far more visible spot in the morning due to higher flow through the pipe at peak water demand. Another important factor skewing results in this image is solar loading of the roadway, which will be explained in the next section.

Solar loading is the effect the sun has of inputting heat energy into all objects that receive sunlight. Solar loading causes a net increase in radiated energy onto objects that may severely skew results in a thermal image.

In Figure 11.5, the same section of asphalt that is exposed to the sun appears to be 10 °F warmer than the section in the shadows. This effect occurs quickly—the parking lot in Figure 11.5 was imaged less than an hour after sunrise. This temperature difference is entirely due to sunlight and is not representative of any other source of heat or anomaly in the scene. The effect can also be seen in even greater measure on the building rooftop in the corner of the image. For an inspector looking to get accurate imaging of anomalies under a roadway or on a rooftop, this would certainly be a poor time of day to take thermal images. For best results, one must closely consider the thermal dynamics of their project and plan time of day accordingly.

FIGURE 11.5   Thermal image of solar loading occurring shortly after sunrise in a parking lot. In just under an hour, sunlight has created a 10 °F (5.56 °C) difference in temperature between the area of the parking lot in sunlight and in the shade.

Source: Reprinted with permission from Harkin Aerial.

Ambient temperature and subject temperatures are especially important factors to consider. To detect heat transfer in thermography, there must be a temperature difference present in the first place. Therefore, a thermographer is typically looking to image when there is the largest difference in temperature (*delta T*) between the object of interest and ambient conditions. There are three modes of heat transfer: *conduction, convection, and radiation.* Each one is a method by which heat may be exchanged between the subject and the environment, and is described next:

Conduction is the transfer of thermal energy between substances, typically from direct contact. If you pull a pan off a hot stove, all the heat felt on the pan would be heat due to conduction from the stove surface to the pan itself. The formula for determining conduction is known as *Fourier's Law of Heat Conduction*, which is described in Equation 11.1.

$$Q/t = (k \times A \times (T_1 - T_2))/L \qquad (11.1)$$

Where
Q = conductive heat transfer (Joules)
t = time (s)
k = thermal conductivity (J/m-s-°C)
A = area (m^2)
$T_1 - T_2 = \Delta T$ (delta T) (°C)
L = conductive path length or thickness (m)

Convection is a transfer of heat via the motion of a moving fluid, whether forced or natural. A computer component being cooled by airflow across it is an example of convective heat transfer. Convection is also a common mode of heat transfer on rooftops when wind is present. The formula for calculating convection is described in Equation 11.2:

$$Q/t = h \times A \times (T_{surface} - T_{fluid}) \qquad (11.2)$$

Where
Q/t = heat flow (Watts)
h = convective heat transfer coefficient (W m^{-2}K^{-1})
A = surface area (m^2)
$T_{surface}$ = temperature of surface (°C)
T_{fluid} = temperature of fluid (°C)

Lastly, *radiation* is the transfer of heat from a solid, liquid, or gas in the form of electromagnetic waves as a direct result of emission and absorption. As previously discussed, while thermal cameras can indirectly observe the effects of conduction and convection, they directly measure radiation of an object. As such, radiative heat transfer is especially important for a thermographer to understand for proper interpretation of imagery. The *Stefan-Boltzmann Law* describes radiative heat transfer and is shown in Equation 11.3:

$$W = \sigma \varepsilon T^4 \qquad (11.3)$$

Where
W = emissive power (Watts)
σ = Stefann-Boltzmann constant (5.67×10^{-8} Watts/m^2 K^4)
ε = emissivity
T = temperature (K)

Using these equations, thermographers can calculate the expected heat loss and determine if the results measured from thermal imagery align with the expected temperatures.

Emissivity of a material is one of the single most important factors to consider when performing thermal imaging. All objects in our universe are under some combination of absorbing, reflecting, and transmitting radiation. Equation 11.4 describes this relationship, which states that the total incoming radiative energy of an object is the total of energy absorbed, reflected, and transmitted.

$$W_{incoming} = W_{absorbed} + W_{reflected} + W_{transmitted} \qquad (11.4)$$

Where
$W_{incoming}$ = total incoming radiative energy (Watts)
$W_{absorbed}$ = absorbed radiative energy (Watts)
$W_{reflected}$ = reflected radiative energy (Watts)
$W_{transmitted}$ = transmitted radiative energy (Watts)

Similarly, we know that energy can be neither created nor destroyed, which means energy is conserved within the radiative process. Therefore, the absorptivity of a material, plus the reflectivity of a material, plus the transmissivity of a material must sum to 1, as demonstrated in Equation 11.5:

$$a + r + t = 1.00 \qquad (11.5)$$

Where
a = absorptivity of a material
r = reflectivity of a material
t = transmissivity of a material

Most objects are opaque; that is, they do not transmit any thermal energy, they only reflect and absorb. Therefore, t = 0 for most objects that would be imaged by a thermal camera. Knowing this, and knowing that absorptivity equals emissivity at objects at thermal equilibrium, we can rewrite Equation 11.5 as:

$$\varepsilon + r = 1.00 \qquad (11.6)$$

Where
ε = emissivity
r = reflectivity

Equation X.6 shows that an object in thermal equilibrium that also reflects zero thermal energy must be a perfect emitter, that is, emissivity must equal 1. This is known as a *blackbody* and is a theoretical perfect emitter of energy. If a blackbody existed in real life, 100% of its thermal energy would be detected by our thermal cameras. Of course, ideal conditions such as this do not exist in real life and emissivity of materials can vary considerably.

Why is emissivity important to understand? Since a thermal camera is based on the concept of measuring emitted radiation, objects at lower emissivity will appear to be a *lower* temperature than objects of higher emissivity in a thermal image, even if they are the same temperature in reality. To obtain correct temperature readings, we must look up the emissivity of a material ahead of time, or measure it independently, and input that value in the thermal camera settings. Lower emissivity materials (such as metal or glass) tend to be highly reflective and are consequently harder to measure than higher emissivity materials (such as soil, brick, or clay). Fortunately, many materials of interest to the thermographer will be of high emissivity, but metals are a common material where special care is needed to determine and account for emissivity and incorrect temperatures from reflections. Modern thermal cameras include post-processing software where these values can be input to compensate and obtain more accurate temperatures of a particular material. However, as aerial thermography can suffer from large reflections of the day or night sky, properly measuring and compensating for emissivity and reflected temperature in such a thermal scene is still a non-trivial task. As a result, UAV thermal imaging is best suited for qualitative studies where finding a general temperature difference is the goal but not quantitative studies where an absolute temperature must be measured and compared to a known value.

In the final section, a case study will be discussed that leverages the unique benefits of both UAVs and thermal imaging in a novel and innovative manner.

CASE STUDY—THERMAL ILLICIT DISCHARGE DETECTION IN MUNICIPAL SEPARATE STORM SEWER SYSTEMS (MS4S)—NEW YORK, USA

In recent years, the New York State Department of Environmental Conservation (NYSDEC) has increased reporting requirements for operators of municipal separate storm sewer systems, also known as MS4s. An MS4 is a conveyance or a system of conveyances owned by a city, town, or public entity that is not connected to a sewage treatment or combined sewer system. The release of potentially harmful pollutants (known as *illicit discharges*) is a major environmental concern for MS4s, and the proper mitigation of pollution into an MS4 has been a pressing issue for NYSDEC and many municipalities throughout the state. In 2020, a pilot study was carried out to determine the efficacy of using thermal camera–equipped UAVs to detect illicit discharge activity. The study, funded by Nassau County, covered several watersheds on the north shore of Long Island where illicit discharges were suspected as contributors to poor water quality and high fecal coliform bacteria measured at local beaches and harbors. By utilizing a FLIR XT2 camera mounted on a DJI M600 hex copter UAV, six locations covering five separate stormwater conveyances were flown within the span of one week, from November 29, 2020, to December 4, 2020. The study acquired over 13,000 thermal images and mapped several miles of coastline and stormwater conveyances. The FLIR XT2 is an uncooled microbolometer thermal camera with 640 x 512 resolution, one of the highest resolutions commercially available in the small-to-mid UAV market. A 13 mm focal length lens was chosen to balance moderate field of view with good thermal spot size ratio, which is a measure of the smallest feature that can be identified on the camera at a given distance.

As previously discussed, many environmental factors can incorrectly skew the results, and a week-long UAV mission such as this required close attention to time of day and weather patterns. December 2020 was chosen for a variety of reasons: in the Northeast United States, this month produces the largest delta T between warm water sources and colder ambient temperature. December is a leaves-off period in the Northeast US, which increased the chances of spotting thermal anomalies through tree cover. To reduce chances of convection skewing results, flights were performed at wind speeds below 5 knots. To eliminate results from solar loading, flights were performed from one hour before sunrise to sunrise or from one hour after sunset to two hours after sunset. Lastly, it was understood by the researchers that materials of varying emissivity would be appear in one image, and as such only *qualitative* apparent temperatures were considered: temperatures were not corrected for emissivity of materials and, therefore, understood to not be absolute readings.

The study produced notable findings, such as the discovery of an abandoned drinking water well discharging into a local bay (Figure 11.6), an overflowing stone well discharging into the same bay, (Figure 11.7) and several discharges from roof or basement drains into nearby storm drains (Figure 11.8, Figure 11.9).

While no single discharge was seen as a primary culprit of local water quality concerns, thermal imaging results agreed with independent water sampling procedures and further lent credence to the prevailing theory of fecal coliform bacteria originating from animal pet waste washout into storm sewers, which would not be easily identified on a thermal camera.

CONCLUSION

Aerial thermography is both a challenging and rewarding field of study, combining two distinct areas of expertise—UAV piloting and thermography—into a single planned mission. As such, it often requires a diverse team with expertise in flight operations, thermal data collection, mapping, imagery post-processing, and data management. As the cost of acquiring UAVs and thermal cameras continues to rapidly drop, this technology is quickly undergoing further democratization and widespread use. While an effort has been made in this chapter to cover both breadth of thermal applications and the depth of a particular use case, the applications of this technology are nearly limitless. The reader is encouraged to seek out additional resources—a wealth of training and research materials are published each year

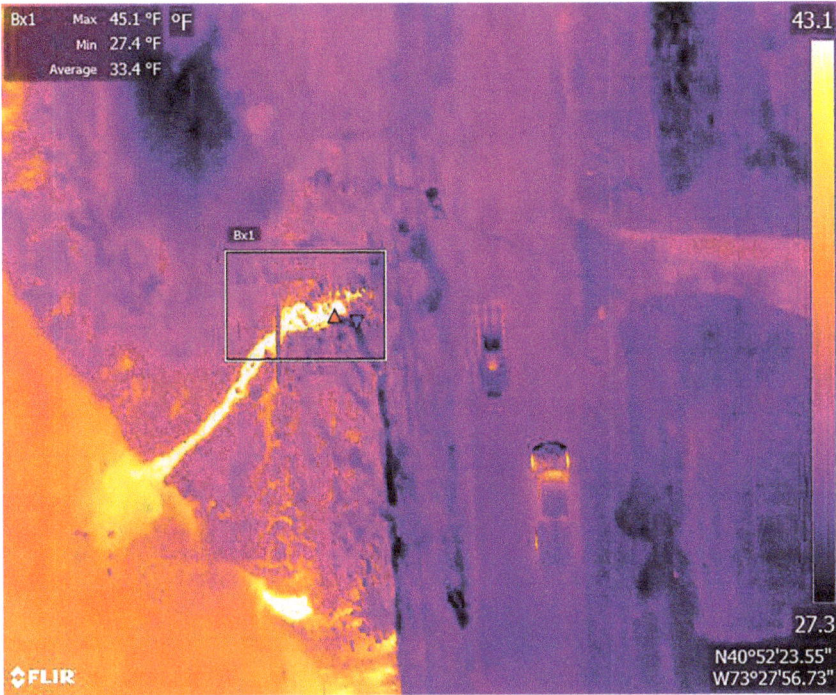

FIGURE 11.6 Thermal image of an abandoned drinking water well discharging into a local bay. Although the well pipe was less than 2 inches in diameter, the flow and size of the discharge was significant enough to create a thermal signature easily seen by a UAV. The discharge was spotted mid-flight by a UAV at approximately 120 feet above ground level.

Source: Reprinted with permission from Harkin Aerial.

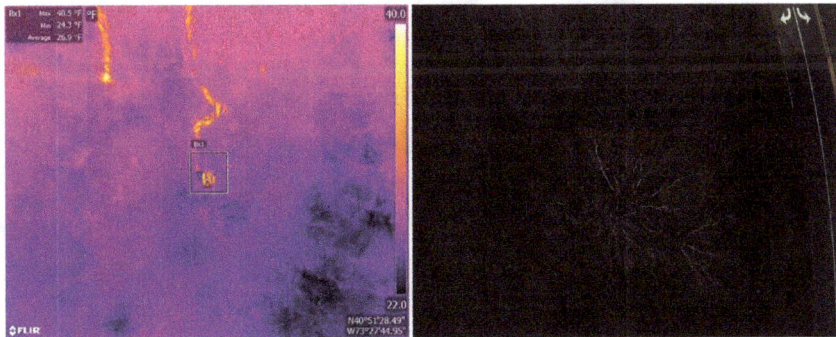

FIGURE 11.7 Thermal image (left) and visible light image (right) of an overflowing stone well. The well is nearly impossible to spot under the tree cover from the visible light image. On the thermal image, both the well and the overflowing water are easily identifiable.

Source: Reprinted with permission from Harkin Aerial.

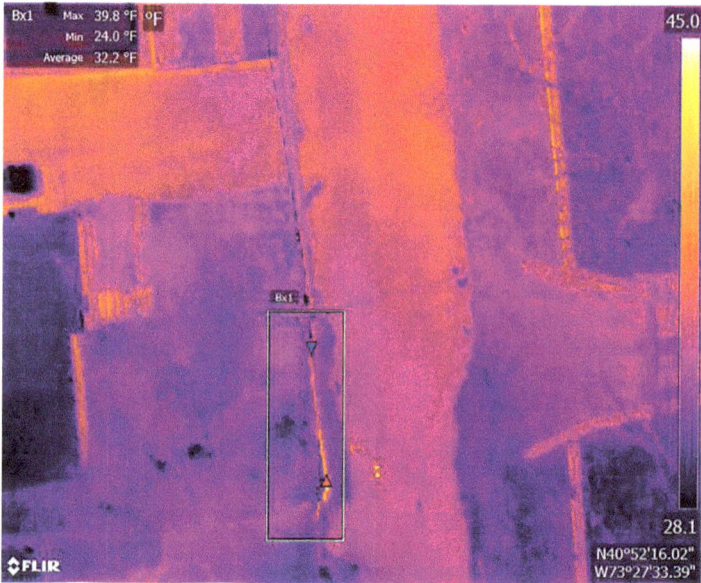

FIGURE 11.8 Thermal image of a discharge from a buried pipe leading out to the roadway. At the time of the study, the pipe had an unknown origin but is likely a basement or roof drain from a nearby commercial building. A ground follow-up revealed recently disturbed pavement and broken piping, suggesting a recent relocation by the property owner.

Source: Reprinted with permission from Harkin Aerial.

FIGURE 11.9 Thermal image of another discharge from a buried pipe, flowing out to a storm drain. Untreated discharges from basement drains are a leading cause of small-scale illicit discharges. The ability to quickly find and geo-tag discharges for ground follow-up makes UAV thermography an excellent tool for environmental monitoring.

Source: Reprinted with permission from Harkin Aerial.

by camera manufacturers such as FLIR[1] and Workswell.[2] Standards such as those published by the American Society for Testing and Materials (ASTM)[3] are also an excellent resource for those interested in learning more about this exciting and ever-changing field.

NOTES

1. www.flir.com
2. www.drone-thermal-camera.com
3. www.astm.org

REFERENCES

"DJI and FLIR Systems Collaborate to DEVELOP Aerial Thermal-Imaging Technology." *DJI*, October 12, 2015. www.dji.com/newsroom/news/dji-and-flir-systems-collaborate-to-develop-aerial-thermal-imaging-technology.

Gromicko, Nick, and John McKenna. "The History of Infrared Thermography." *InterNACHI*. Accessed August 3, 2021. www.nachi.org/history-ir.htm.

Karpowicz, Jeremiah. "Detailing the Success of the L.A. Fire Department's Drone Program." *Commercial UAV News*. Commercial UAV News, November 28, 2018. www.commer cialuavnews.com/public-safety/success-lafd-drone-program.

Westervelt, Jason R., H. Abarbanel, R. Garwin, R. Jeanloz, J. Kimble, J. Sullivan, and E. Williams. *JSR-97–600 Imaging Infrared Detectors II*, 2000. https://irp.fas.org/agency/dod/jason/iird.pdf.

White, Jack R. "Herschel and the Puzzle of Infrared." *American Scientist*, January 9, 2019. Accessed August 3, 2021. www.americanscientist.org/article/herschel-and-the-puzzle-of-infrared.

12 Reconstruction of the Ballintaggart Court Tomb Using 3D Scanning, 3D Printing, and Augmented Reality (AR)

*John Meneely, Colm Donnelly, Ciaran Lavelle,
Tony Martin, Brian Sloan, and Stephen Weir*

HISTORY OF THE COURT TOMB

The megalithic monument from Ballintaggart (Historic Environment Record of Northern Ireland (HERoNI) Sites and Monuments Record Number ARM 009:006; Grid Reference H9742052130) was a court tomb located in the townland of the same name in the civil parish of Kilmore, County Armagh, Northern Ireland. One of the four types of a megalithic tomb to be found in Ireland (and also termed as horned cairns, court cairns, chambered graves, chambered cairns, or forecourt cairns in the archaeological literature), court tombs comprise:

> a stone-built gallery consisting of between two and five chambers, usually separated from each other by stone slabs. The feature that gives this type of monument its name is the 'court', which consists of a row of upright stones at the entrance to the gallery . . . usually arranged in the shape of a semi-circle.
>
> (Welsh 2011, 41–43)

Typically the gallery would be roofed by stone slabs and the tomb was then covered by a cairn of stone and earth. There are 126 known court tombs in Northern Ireland and some 420 throughout the island of Ireland; radiocarbon dating suggests that they were in use during the Neolithic, between 3720 BC and 3560 BC (Welsh 2011), and were used as burial sites by the first farming communities.

The tomb's original location in Ballintaggart is not marked on the first edition Ordnance Survey six-inch mapsheet of c. 1830, nor those of the second edition (c.1860) or third edition (c.1900), although there appears to be a tree-ring or landscape feature at its location, perhaps associated with nearby Ballintaggart House. It is only with the fourth edition from 1905 that the monument is marked, denoted by the word "Grave," but with a quarry now evidently active. By the time of the

 DOI: 10.1201/9780429327575-12

1:10,000-metre mapsheet of the 1950s, the tomb is set in the middle of the quarry, labelled in a Gothic script as a "Chambered Grave" with quarry buildings surrounding it (Figure 12.1). A second megalith (ARM 009:026), also possibly a court tomb, is reputed to have been located close-by but was reported as near obliterated in 1891 (Dugan 1891, 488); we can only assume that its location has now been consumed by the quarry, itself now a small lake.

The Ballintaggart tomb is included in the *Preliminary Survey of the Ancient Monuments of Northern Ireland* (1940) with a short account supplied by Thomas Patterson (1888–1971), the first curator of Armagh County Museum, in which he noted that the site was "enclosed in a modern wall and planted with trees". He also stated that the court's façade was "unusually flat in plan and consists of single standing stones set on each side of the portal uprights" (PSAMNI 1940, 63). Patterson also noted that the site had been excavated in 1902 by Miss Mary Bredon of Ballintaggart House and that this work furnished a sherd (presumably of pottery) "and many white quartz stones" (*ibid.*). De Valera (1959–1960, 124) included the site in his study of the court tombs of Ireland where he noted that it was "well preserved. Mound obscured by fences. Fine four-chambered gallery. Two court-stones flank the entrance. Good segmenting jambs". An account of the tomb, with a plan, was also included in the *Archaeological Survey of County Armagh* (Neill 2009, 102).

The extension of Troughton's Quarry in 1962 was sufficient for Dudley Waterman (1917–1979), Principal Inspector of Historic Monuments with the Archaeological Survey in the Ministry of Finance, to note that this was "cause for disquiet" given that "the quarry now extends on the west right up to the ditch which surrounds the remaining stones of the gallery and the forecourt". It would not be for a further four years, however, before action would be taken to resolve the long-term future of the tomb, commencing with an excavation in 1966 by Laurence Flanagan (1933–2001), Keeper of Antiquities in the Ulster Museum, in advance of its deconstruction

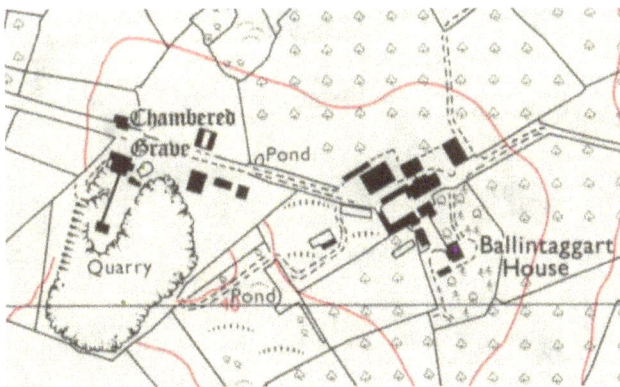

FIGURE 12.1   The 1950 1:10,000 m map showing the location of the tomb marked as a 'Chambered Grave' and near-surrounded by the encroaching quarry that threatened its existence.

and removal to a safer environment. While the results of Flanagan's work were not brought to final publication, he did compile a helpful preliminary report, accessible within the monument's HERoNI file (Figure 12.2).

From this report it can be established that only a small part of the tomb's cairn remained in place and that the gallery had been some 13.5 metres in length (east-west) and 5 metres in width (north-south). The forecourt was at the eastern end, with four upright stones flanking entrance into the gallery, divided into four chambers by the use of jamb stones. The excavation revealed that the four chambers had been disturbed, presumably during Miss Bredon's excavation of 1902. As noted previously, this work retrieved a sherd of pottery and many white quartz stones (PSAMNI 1940, 63) and Flanagan (1966) noted that the four chambers "had suffered disturbance; indeed locally it was remembered that a certain Charlie Doherty, who died 15 years ago, had dug them 'to the floor' for Miss Bredon". Very few artefacts were recovered, as a consequence, by Flanagan with the exception of Chamber 3, where a small undisturbed area was encountered that proved to contain over 100 flakes of flint, an end scraper, and a core. Excavation in the forecourt area, however, produced quantities of Neolithic pottery (Western Neolithic/Lyles Hill type carinated bowls), suggesting that this area had not been investigated by Bredon. The letter from 1962 contained within the site's HERoNI file indicates that it was Waterman's objective that the tomb is dismantled and removed for re-erection at a

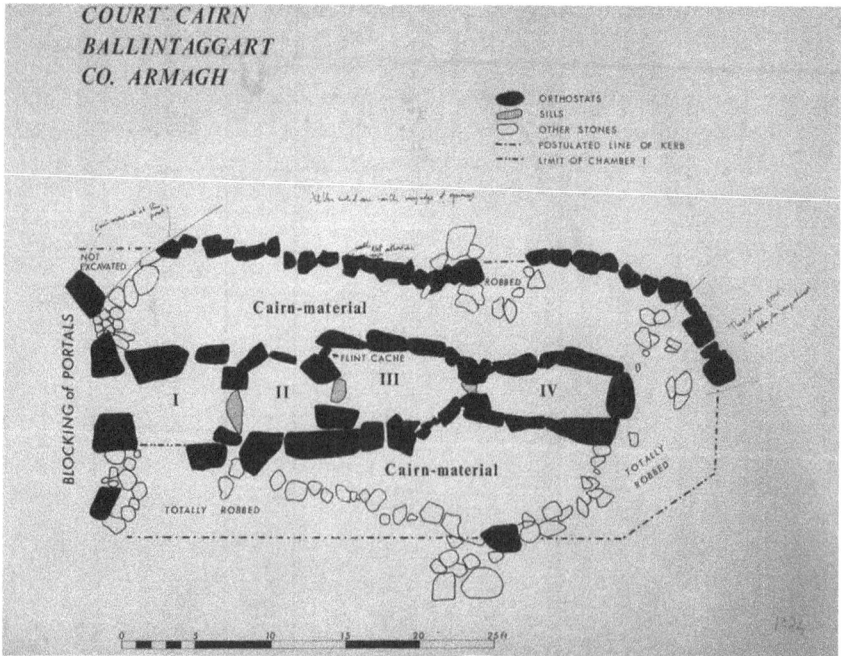

FIGURE 12.2   The plan of the tomb produced by Flanagan during the course of the excavation in 1966.

suitable safe location. As such, in 1966 Flanagan produced a detailed plan of the monument that would enable it to be reconstructed as accurately as possible at a new home.

Temporarily stored in Hillsborough, it was decided that this should be beside the refurbished and expanded Ulster Museum in Botanic Gardens in south Belfast, where it became a popular visitor attraction and one that was used year-on-year by staff from the archaeology department at Queen's University Belfast as a survey-ing training facility for undergraduate students (Figure 12.3). In 2006, however, a new phase of refurbishment at the museum necessitated the dismantling of the tomb for safe storage at the National Museums Northern Ireland (NMNI) prop-erty at Cultra, County Down. In an effort to find a new and safe location for the tomb, the museum staff commenced a consultation process with stakeholders in 2019, which determined that within the grounds of the Ulster Folk Museum would be the best solution, providing a link between our archaeological heritage and the folk history represented at the museum. When Waterman had first pondered the removal of the tomb in 1962, he had stated that an ideal place for its long-term future would be the Ulster Folk Museum. Given this, it is apt that this is, indeed, now where the tomb will be relocated and where it can be enjoyed by future generations.

FIGURE 12.3 Archaeology students from Queen's University Belfast undergoing survey training at the tomb when it was located at the Ulster Museum in Botanic Gardens, Belfast (2003).

Prior to the redevelopment of the Ulster Museum in 2006, the megalith stones—which ranged in size from a few kilograms to several tonnes—were removed and stored outside on wooden pallets in a storage yard at the Ulster Folk Museum. With the decision made to reconstruct the tomb in the grounds of the folk museum, the museum's curators made contact with staff within the School of Natural and Built Environment at Queen's University Belfast with a view to the establishment of a specialist team drawn from geography, archaeology, and civil engineering that might help ensure the reconstruction was as accurate as possible, with new construction plans created that were based on drawings, measurements, photographs, and other historic records of the tomb. Archival investigations to retrieve such materials, however, met with limited success, and it was concluded that the existing collection of stones was essentially a '3D jigsaw' that would require a degree of interpretation on behalf of the project team if it was to arrive at an authentic arrangement for the reconstructed tomb. This would necessitate some 'trial and error' with the stones to establish an optimum final arrangement. However, from a visual inspection of the existing stones, it was evident that frost damage had occurred to some and that they were relatively fragile. It was, therefore, determined that this approach to the reconstruction using the original stones would risk causing further damage to them and that measures would be needed to allow the archaeological team to test a series of arrangements whilst minimising the risk of damage to the stones themselves. Hence, the decision to 3D scan each stone and produce 1:10 scale replicas was taken.

3D MODELLING THE ORTHOSTATS

As part of the development of the NMNI's digitization infrastructure, investment in cutting-edge scanning technologies resulted in the purchase of an Artec Leo Handheld 3D scanner. The first large project this new digital team undertook was the 3D scanning of the large, individual stones of the deconstructed tomb to aid its physical reconstruction.

The Artec Leo 3D Scanner has an onboard NVIDIA GPU for real-time processing. This processor powers the scanner's built-in scanning software, so it is not necessary to connect it to a computer in order to collect data, making it extremely easy to manoeuvre around objects. The scans can be viewed in real-time on an adjustable touchscreen display. It is powered by a swappable lithium-ion battery, and for post-processing on a PC, the 3D scan data can be transferred via Wi-Fi.

This is a structured-light scanner, which operates by projecting a known pattern of light onto a surface and analysing how it is deformed by that surface to calculate the geometry of an object. When the scanner has projected a structured-light pattern and calculated the geometry, it creates a 'frame' and the Artec Leo scans at 80 frames per second. To register (join together) these individual frames, most structured-light scanners use specially placed targets as a known reference point between frames, and these must be placed on the surface of the object prior to scanning. However, this is a target-less scanner, and it uses a combination of surface geometry and texture (colour) data to register different frames. It should be noted that if this structured-light technology is being used to capture transparent or reflective materials, these can be difficult to capture and may, if possible, need a coat of scanning spray to successfully capture them.

There are two important factors to consider when using a structured-light scanner: the distance from and the angle of the scanner to the object being captured. The best results are obtained when an even distance is kept between the scanner and the object, and for the Artec Leo, this is approximately one metre. This scanner includes a colour distance overlay that displays in real-time if the operator is maintaining the ideal distance—green is good, red is too close, and blue is too far away from the object (Figure 12.4). As for angle, the best results are obtained when the scanner is held normal (90°) to the surface of the object.

In addition to the distance overlay, there is also a quality overlay that can be selected. This registers as green on areas of the object that have been satisfactorily scanned and red on areas that have not. Toggling between these two, distance and quality overlays is good practice while scanning, as it will show how the scan is progressing and when it is complete. Once the scanner is turned on, it is simply a matter of pulling the trigger once to open a scan preview on the display and pulling it again starts the scanning process. It is now a case of carefully moving around the object to build up a 3D scan. It is important to keep the object in the centre of the touchscreen display—if tracking is lost on the object, the scan will automatically stop and it must be re-centred before scanning can recommence. The scanner weighs 2.5 kg (5.4 lbs), which does seem light, but arm fatigue can set in when scanning for long periods. Fortunately, there is a pause and resume function, which, as long as the object being scanned is not moved during its application, can be used to take a break.

FIGURE 12.4   This is an image of Stone 15 from the deconstructed tomb being scanned with the Artec Leo Handheld 3D scanner in distance mode. The green-coloured scan data on the screen of the scanner shows that the data is being collected at the optimal distance.

The individual stones from the tomb were kept on pallets in an exterior area of the Ulster Folk Museum. The first step was to consult with the Archaeological, Collections Care, and Conservation and Collections Information Teams within the museum to identify which stones required scanning and to assess whether they were in an appropriate condition to undergo the process. Thirty-four orthostats were identified and marked out as suitable candidates.

Each stone was then numbered with chalk, and scanning was undertaken when the weather was dry but overcast (Figure 12.4). The stones were scanned using high-definition (HD) mode, which can produce scans with a resolution of up to 0.2 mm. The top surface and sides of each stone were captured and labelled 1A, 2A, 3A, etc, and then the museum's Collections Operations Technical Team was charged with the task of carefully rotating these large, heavy, fragile stones so that their hidden bases from the first scan and a repeat scan of the sides could be captured—these were labelled 1B, 2B, 3B, etc.

After scanning was complete, the data was post-processed on a PC using Artec Studio 15 Professional software. This involved an initial manual cleaning phase of each scan to remove any background data (e.g., parts of the pallet that the stones rested on that had been captured during scanning) (Figure 12.4). The two scans captured on each stone were then manually aligned, using the duplicated side data as a guide for each stone. The autopilot feature in the software was used to accurately auto-align the two scans, create a meshed surface, check for holes, and fill them to make the model manifold before finally adding the texture. The resulting, fully registered, textured scans for each stone were then exported as a .obj (object) file at 1:1 scale.

PRODUCTION OF A SCALED REPLICA OF THE PROPOSED RECONSTRUCTION

The .obj files were opened in 3D Builder, a free, easy-to-use software from Microsoft for designing and preparing objects for 3D printing. This software has an embossing function that enables the user to add text to a 3D part. Each stone was opened separately in this software and embossed with its scan number and then exported as a .stl file at 1:1 scale. This is the most common and universal file format used for 3D printing—STL stands for stereolithography. It is a 3D rendering with a single colour, whereas .obj files store geometry, colour, texture, and material information. These .stl files were then opened in the slicing software Prusa Slicer and scaled to 10% of their original size. Slicing software is used for the conversion of a 3D model, usually in .stl format, to G-code (Geometric Code) for 3D printing in a fused deposition modelling (FDM) printer. The slicer divides the 3D model into a stack of flat layers and then converts these layers into a series of linear movements, along with some printer-specific commands—producing a G-code file. This G-code file is loaded onto the 3D printer for printing. The printer used for this project was a Prusa i3 Mk3S, which has a print volume of 11,025 cm³ (250 mm wide x 210 mm deep x 210 mm high or 9.84 in x 8.3 in x 8.3 in). As more than one stone can be printed at a time at this 1:10 scale on this 3D printer—stones were put into groups that had a total print time not exceeding 36 hours, and it took just over 15 consecutive days to print all 34 stones at 1:10 scale and a further day to remove the support material from each stone.

In parallel with the investigations into the condition and arrangement of the megaliths, an inspection was also carried out into the existing ground conditions at the proposed new site for the court tomb reconstruction within the grounds of the Ulster Folk Museum at Cultra, Co. Down. This involved carrying out a series of percussive borehole investigations down to a depth of 3 metres to obtain representative undisturbed samples of the underlying ground strata. Standard penetration tests were carried out on these borehole samples at regular intervals. Results showed that the site had no previous evidence of ground disturbance and that the underlying soil was mainly composed of stiff gravelly clay and that the water table was below the depth investigated. Hence, the proposed location for the reconstruction was considered to be ideal as the ground bearing capacity (estimated as 100 kPa) was such that it would inhibit any future subsidence of the reconstructed court tomb.

Having reproduced 34 1:10 scaled replicas of the orthostats and having determined that stiff gravelly clay was present at the proposed new site, it was then possible to simulate these site conditions using kaolin clay as a part of a miniature 'set' to allow the archaeological team who were familiar with the court tomb site when it was at the Ulster Museum in South Belfast and using Flanagan's 1966 plan (Figure 12.2) to reconstruct the tomb at a 1:10 scale. Following several trials, the final court tomb reconstruction arrangement was established as a 1:10 scale model that was subsequently 3D modelled using the structure from motion (SfM) software Reality Capture (RC). This involved taking 553 12 MB digital images of the scale model using a Sony A7RIV digital camera with a 50 mm lens set to f5. These images were processed using Reality Capture, a photogrammetry software application for Windows that produces 3D models from a set of overlapping images, and then scaled to 1:1 using measurements taken from the boards that it was created on. The resulting model is shown in Figure 12.5.

FIGURE 12.5   This is an image of the 1:10 scaled 3D model created by the archaeologists.

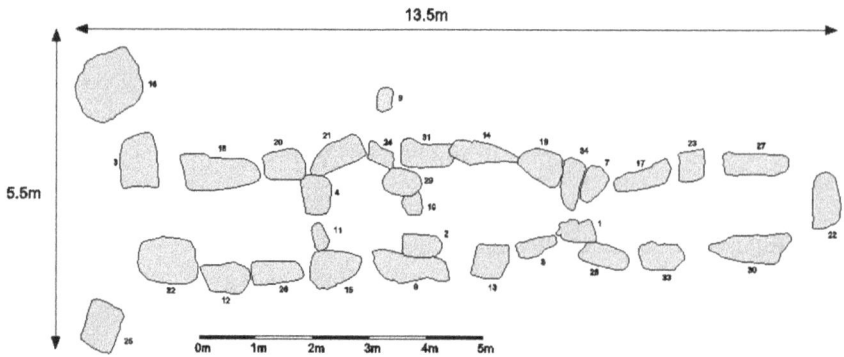

FIGURE 12.6 A 2D plan created by importing the 1:10 scale model into a CAD package and tracing the outline of each stone.

This 3D model of the 1:10 scale tomb was then exported from RC as a point cloud in .xyz format and converted to a ReCap file (.rcp) using Autodesk ReCap Pro software for import into Autodesk AutoCAD. The point cloud data was then scaled up to 1:1 (x10 the imported model) and the outline of each stone, in orthographic top view, was then traced and numbered to produce a simplified 2D plan of the proposed reconstruction (Figure 12.6)

The completed 3D model was then uploaded to Sketchfab, a web-based platform used for sharing/viewing 3D models. This model is available to view at the following link: https://sketchfab.com/1manscan/models. One advantage of this website is the ability to view the model of the reconstructed tomb in augmented reality (AR). As the model was uploaded at a 1:1 scale of the tomb, this AR facility can be used to ensure that the actual megaliths are in the correct place/orientation by simply viewing and comparing the model in AR through a smartphone or tablet on-site while the stones are being placed. This is an extremely powerful yet simple tool that will be an invaluable aid during the tomb's reconstruction, ensuring that it accurately represents the archaeological team's modelled reconstruction.

CONCLUSION

The megalithic tomb from Ballintaggart has had a varied story over the past 60 years, from being on the edge of total destruction by quarrying at its original location, its excavation, and dismantlement in 1966 and removal to the Ulster Museum, to its second dismantlement in 2006 and subsequent storage at the Ulster Folk Museum. It remains, however, a valued component within the museum's collection, and all this preliminary 3D work conducted by the team from Queen's University Belfast will vastly reduce the time and effort on-site in correctly placing these delicate stones and enable the museum curators to advance with their plans to display the tomb as an outdoor exhibit. It will be a fitting reminder to visitors at the Folk Museum of our Neolithic ancestors who first farmed this land.

REFERENCES

De Valera, R., 1959–1960: "The Court Cairns of Ireland", *Proceedings of the Royal Irish Academy* 60C, 9–140.

Dugan, C.W., 1891: "Cromlech Near Portadown", *Journal of the Royal Society of Antiquaries of Ireland* 1, 488.

Flanagan, L., 1966: *Excavations at Ballintaggart, County Armagh: Preliminary Report*, Unpublished Excavation Account Contained within the HERoNI file for ARM 009:006.

Neill, K., 2009: *Archaeological Survey of County Armagh*, Northern Ireland Environment Agency, The Stationery Office, Belfast.

PSAMNI 1940: *A Preliminary Survey of the Ancient Monuments of Northern Ireland*, His Majesty's Stationery Office, Belfast.

Welsh, H., 2011: *Tomb Travel: A Guide to Northern Ireland's Megalithic Monuments*, Publisher Stationery Office, Norwich, UK.

13 Terrestrial Laser Scanning for Monitoring and Modelling Coastal Dune Morphodynamics

Sarah Kandrot

INTRODUCTION

Quantitative information on coastal topography is often an essential component of geomorphic inquiry. This information can be obtained in the field using a theodo-lite, electronic distance meter (EDM), precise Global Navigation Satellite Systems (GNSS), or other similar technologies. To obtain elevation data with these tools, usu-ally single regularly or irregularly spaced elevation measurements are taken along a linear transect. This approach to surveying is limited, though, in that, in practice, only a small number of data points can be captured in the time it takes to complete a typical field survey. Advances associated with airborne and satellite altimetry tech-nologies in the 20th century have made it possible to more efficiently collect many elevation measurements (on the order of millions or hundreds of millions) over a

DOI: 10.1201/9780429327575-13

large area in a short period of time. This technology has been adopted by geomorphologists to map coastal features and has led to improved knowledge of coastal geomorphic processes (Brock and Purkis 2009).

The more recent advent of light detection and ranging (LiDAR) technology has further improved our ability to map smaller scale geomorphic features and processes. LiDAR is an active remote sensing technology that uses either a reflected laser pulse or differences in phase from a continuous beam to measure the distance to a surface. Pulse-based sensors sweep millions of laser pulses across a surface and use the time it takes for those pulses to be reflected back to the instrument to measure the distance to the surface. A rotating optical mirror directs the pulses over the area to be surveyed at spatial intervals specified by the user. There are two general types of LiDAR systems—airborne and ground-based. Airborne systems are flown on an aircraft and, thus, are capable of capturing data over a relatively wide area. They consist of three main parts: the sensor, the inertial measurement unit (IMU), and the GNSS, which work together to produce georeferenced topographic data. Ground-based LiDAR systems, or terrestrial laser scanners (TLS), capture data from one or more fixed positions on the ground. Georeferencing is usually established through the use of a known benchmark, although newer models feature a built-in GNSS. The result of a LiDAR survey is millions of densely-packed 3D points, each with a unique x,y,z coordinate, collectively known as a point cloud (Figure 13.1).

Although ground-based LiDAR systems are limited in terms of coverage area compared to airborne systems, the operational costs associated with these systems are much lower and the instruments can more easily be deployed on demand and at short notice. Ground-based laser scanners are, therefore, often better suited to the study of beach response resulting from short-term forcings, such as storms, aeolian processes, and seasonal changes in wave climate, particularly when data from

FIGURE 13.1 Point cloud showing beach and foredunes (centre) at Rossbehy, Co. Kerry. The resolution of the cloud is approximately 1 cm, decreasing with distance from the instrument. The height of the dune scarp (right of centre) is approximately 4 m.

multiple surveys are required. TLS systems are also advantageous over aerial systems in the study of micro-scale morphologies because they can capture higher resolution data, including for near vertical surfaces. For example, terrestrial laser scanners are capable of capturing point densities three orders of magnitude greater than airborne systems, with many instruments able to capture sub-centimetre resolutions.

This chapter describes the use of TLS to monitor coastal dune morphodynamics as part of PhD research on the impacts of storms on a coastal barrier system on the southwest coast of Ireland (Figure 13.2). A two-year monitoring campaign was undertaken at the Inch and Rossbehy barriers, located in Dingle Bay, County Kerry, Ireland, from May 2012 to July 2014. The purpose of this campaign was to collect information about how the foredunes responded to storm events in an effort to evaluate the importance of the storm driver.

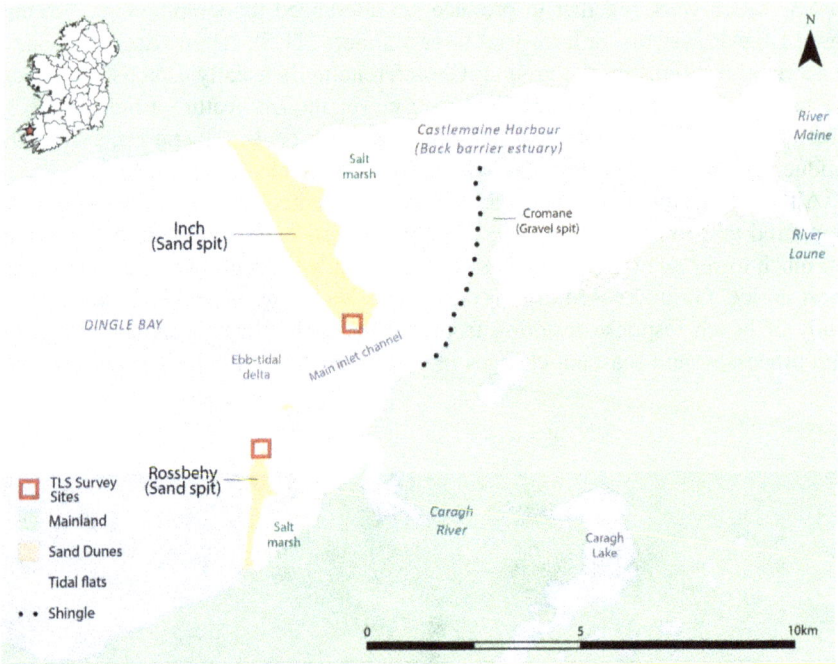

FIGURE 13.2 TLS field surveys were undertaken at Inch and Rossbehy beaches on the southwest coast of Ireland. The barriers are important sediment stores and form part of an extensive and dynamic ebb tidal delta. The Rossbehy spit-barrier has undergone significant changes since breaching occurred in 2008. Prior to this, the high (10+ m) barrier dunes extended north to the area that is now a small island (shown in the map). The breach continued to widen over the course of the study period (from 2012–2014), with the barrier dunes receding landward by more than 90 m. Massive volumes of sediment were redistributed seaward on the tidal flats. Some of the sediment may be returned to the dunes during fair-weather conditions, but the net effect of dune erosion caused by storms has been a shoreline retreat. No such significant changes were observed at Inch. Both sites are important for tourism and recreational activities and are part of Ireland's *Wild Atlantic Way* tourism route.

The data was also used to evaluate the effectiveness of a hydrodynamic model to simulate sediment movements at the site. TLS was chosen as the basis for data collection. The tool was chosen for a number of reasons. Firstly, with TLS, topographic data can be captured on demand. This was a requirement of this research in that to understand the response of the dunes to storms, surveys must be completed in the direct aftermath of such events. Similar technology, such as aerial LiDAR, would have been impractical for this research in that surveys can rarely be completed on demand, and even to perform a single survey would have been out of the budget of this study. Second, TLS technology is better able to capture the spatial heterogeneity of surveyed topography than more widely used techniques, such as differential or RTK GPS or electronic distance meter (EDM) surveying because millions of measurements can be captured relatively quickly (as opposed to tens or hundreds with GPS or EDM surveying). This meant sediment volume changes could be calculated with more precision and smaller changes in volume could be detected. Finally, the laser scanner was freely available and provided for this research by the Geography Department at University College Cork.

The overall objective of the laser-scanning campaign was to generate bare-earth digital elevation models (DEMs) and elevation and volume change maps. The workflow described in this chapter is summarized in Figure 13.3. Issues addressed in this chapter include registration of multi-temporal scans, vegetation filtration, generation of DEMs, and quantification of topographic and volumetric change.

TLS DATA COLLECTION

TLS surveys were conducted at the Inch and Rossbehy barriers in Dingle Bay, County Kerry from May 2012 to July 2014. The sites surveyed each covered an area of approximately 100 m x 50 m. The Inch site lies adjacent to the main tidal inlet that separates the two barriers. Scans here covered a large section of the face of a high but gently sloping foredune (>20 m ODM), an extensive ephemeral embryo dune field and part of the upper beach. The Rossbehy site is located at the terminus of the main section of the Rossbehy barrier, adjacent to a newly formed tidal inlet. Scans here mainly covered the upper beach and foredune scarp on the seaward side of the barrier, although some scans covered part of the vegetated dune field behind the scarp and the upper back barrier beach fronting these dunes.

Topographic field surveys at each site were completed using a Leica ScanStation, a pulsed laser scanner with a positional accuracy of ±6 mm up to a range of 50 m (Leica Geosystems 2006). Features of the ScanStation include a pulsed proprietary microchip laser (Class 3R, IEC 60825–1), an optomechanical mirror system with a full 360° horizontal and 270° vertical field of view, and an integrated high-resolution digital camera. The setup primarily consists of the scanner itself, a tripod, a battery pack, and a laptop from which the instrument can be operated. During the course of the monitoring period, problems related to battery performance necessitated the use of a generator, rather than a battery pack, to power the instrument. A Leica C10, a newer and more efficient model than the ScanStation, was used for some of the surveys due to technical difficulties with the ScanStation.

As neither field site was accessible by car, alternative arrangements for transporting the equipment to and from each site had to be made. At Rossbehy, field

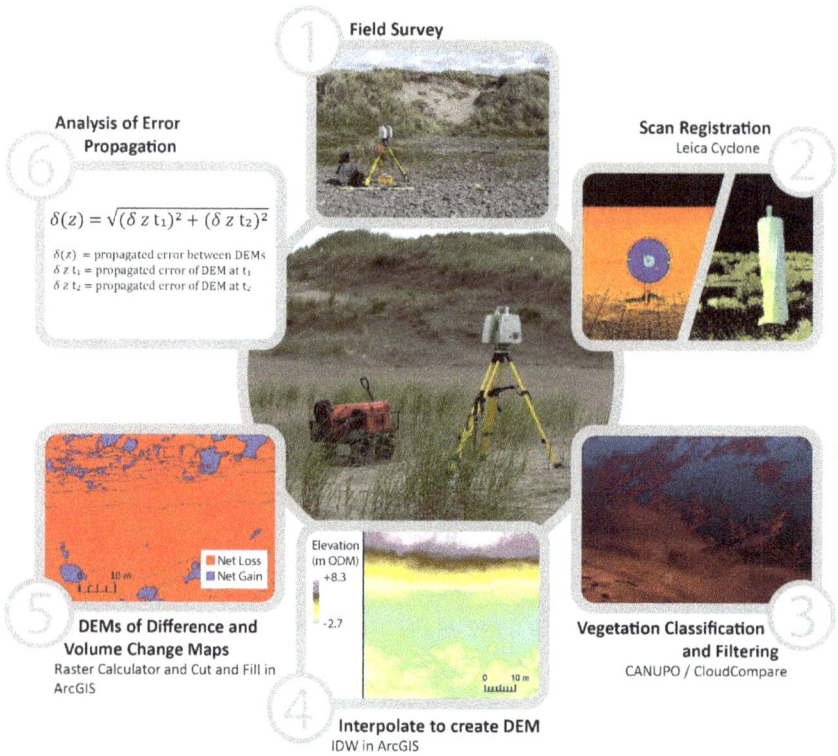

FIGURE 13.3 Workflow for generating bare-earth DEMs and elevation and volume change maps from TLS data. Two instruments were used for this research—the Leica ScanStation (top) and the Leica C10 (centre). The key software packages used for data processing and analysis were Leica Cyclone, CloudCompare, and ESRI ArcGIS.

equipment was transported on a trolley by foot from the nearby car park to the field site, a distance of approximately 2.5 km. At Inch, this distance was considerably further—approx. 5.5 km—so arrangements were made with local farmers to tow the equipment using a quad bike or tractor. Both sites were often inaccessible at and around high tide (MHWS = +3.76 m ODM); therefore, careful planning (and common sense) was required to ensure safe passage on entry and return. Sometimes it was necessary to compromise on the planned coverage area, number of scans, or scan resolution to reduce scan time in the field, particularly on days when there was inclement weather or any concern that it would be possible to return safely. Field surveys were typically carried out as follows. Approximately 15 minutes was required for instrument setup (levelling, powering on, setting up laptop) at each scan station. Leica high-definition surveying (HDS) registration targets were set up strategically such that they would each be visible from each station for subsequent registration. The laser scanner itself was controlled from a laptop using Leica Cyclone v. 8.1, Leica's 3D point-cloud processing software. In Cyclone, the scanner would first be directed to take photos in 360° space around the scanner to aid in selecting the area to be scanned. From the resulting photo mosaic, the three HDS targets, along with semi-permanent targets for multitemporal scan registration, could be identified. Once visually identified, the registration markers are 'fenced'—their locations in 360° space are specified to the software by drawing a perimeter around each one in the photo mosaic. The targets were then scanned in high resolution (1 mm), providing precisely located markers required for later scan registration. Once this was complete, the area to be scanned was fenced. Once the resolution and range were specified, the scanner could be directed to begin scanning. A typical scan for a 180° scene with 2.5 cm resolution over a 30 m range would take approximately 1.5 hours with the Leica ScanStation. Scans were obtained at resolutions of 1–15 cm. The desired optimal scan resolution of 1 cm was based on previous studies that used TLS to study seasonal variations in dune morphology (e.g. Montreuil et al. 2013; Feagin et al. 2012). In practice, it was difficult to obtain 1 cm resolution scans for this study due to limitations associated with the efficiency of the instrumentation and tidal and weather conditions. In terms of time intervals between surveys, an

TABLE 13.1
Summary of TLS Scan Data Obtained during Surveys Completed at Field Sites

Site	Date	Instrument	Resolution (cm)	Total Number of Points in Cloud
Inch	2012–05–24	Leica ScanStation	1.00	8,367,215
Rossbehy	2012–06–28	Leica C10	2.00	67,420,725
Rossbehy	2012–08–05	Leica C10	2.50	97,274,308
Inch	2012–08–06	Leica ScanStation	2.50	30,972,308
Inch	2012–10–06	Leica ScanStation	2.50	47,232,935
Rossbehy	2012–10–07	Leica ScanStation	2.50	43,358,639
Rossbehy	2012–11–15	Leica ScanStation	2.50	4,267,504
Inch	2013–01–09	Leica ScanStation	2.50	10,058,134
Rossbehy	2013–01–30	Leica ScanStation	2.50	4,699,073
Inch	2013–02–27	Leica ScanStation	2.50	4,843,043
Rossbehy	2013–02–28	Leica ScanStation	2.50	7,609,265
Rossbehy	2013–04–19	Leica ScanStation	2.00	7,459,604
Inch	2013–05–02	Leica ScanStation	2.50	7,804,177
Rossbehy	2013–06–05	Leica ScanStation	2.50	6,794,554
Inch	2013–06–20	Leica ScanStation	2.50	24,366,526
Rossbehy	2013–08–06	Leica ScanStation	2.50	5,023,912
Rossbehy	2013–12–11	Leica ScanStation	10.00	4,104,385
Rossbehy	2014–01–16	Leica ScanStation	15.00	1,573,813
Inch	2014–03–12	Leica ScanStation	2.50	23,432,492
Rossbehy	2014–05–04	Leica ScanStation	2.50	22,997,536
Rossbehy	2014–07–29	Leica ScanStation	2.50	16,386,541
Inch	2014–08–28	Leica ScanStation	2.50	2,358,641

approximately 1–2 month interval was desirable such that morphological changes could be attributed to particular events. Overall, 22 surveys were completed—9 at Inch and 13 at Rossbehy. Table 13.1 summarises information about the data collected during the individual field surveys.

The data captured by the scanner for each point included the x, y, and z coordinates, laser-scanned intensity values, and RGB values obtained from photographs taken with the built-in digital camera. The x,y,z coordinates were collected relative to the position of the scanner in an arbitrary grid reference system. In this system, the scanner is located at the (0,0,0) coordinate. Each point in the cloud is represented by its distance in the x, y, and z directions (in metres) from the scanner. Laser scanned intensity is also recorded by the scanner and represented In Cyclone on a scale ranging from—2048 to + 2048. Because each individual scanner can have different intensity characteristics, the values supplied by the scanner to Cyclone are scaled to this range. RGB values are given from 0–255 as per the RGB colour model. The data are stored in the Cyclone database as project (.imp) files and can be exported in a variety of formats, including as ASCII text.

POST-PROCESSING

Post-processing of the data involved (1) scan registration, (2) vegetation classification and filtration, (3) DEM generation, and (4) chronotopographic and volumetric change analysis. These steps involved testing different methods to identify the most appropriate ones for the specific data sets being modelled. This section focuses on those that were found to work best.

SCAN REGISTRATION

The reference datum for each TLS point cloud is the position of the scanner; therefore, all clouds are collected in arbitrary coordinate systems and must be registered to one another. Two general types of scan registration were performed for this research—same-date scan registration and different-date scan registration. Scans obtained from two or more stations on same-date field surveys were each registered to a single arbitrary common coordinate system using the Leica HDS registration targets. A minimum of three Leica HDS targets were set up in the field such that they were visible to the scanner from all stations. The position of the targets is important, as the RMS error associated with registration can only be guaranteed for the area enclosed by the targets, so the wider the area enclosed by the targets, the better. The targets were manually defined in Cyclone in the field (e.g., named t1, t2, t3) and a fine scan (1 mm) of each target was completed at each station. Corresponding target names were used to register the cloud obtained at Station 2 to the cloud obtained at Station 1. The resulting registered (or 'unified') cloud is, therefore, in the coordinate system of the cloud obtained at Station 1. This was performed for each of the same-date scans in Cyclone. Cyclone provides a report on the overall accuracy of the registration. This includes the error in the x, y, and z directions and the root mean square (RMS) error for each target constraint. RMS errors between same-date scans were typically around 0.01 m. The maximum error of registration was 0.025 m.

Different-date scans also had to be registered to one another so that elevation and volume change between them could be calculated. This was done using a differential GPS and stationary, semi-permanent markers installed in the field—wooden posts with steel nail heads protruding approximately 5 cm out of the tops of the posts. The nail heads acted as registration markers and were easily identifiable within the scans. While these registration markers stayed in place for the duration of the study at the Inch field site, they were removed on three occasions from the Rossbehy field site due to rapid erosion and possible theft. Because the GPS coordinates of the posts were taken each time with a differential GPS (dGPS) as a precautionary measure, scans lacking common coordinates from semi-permanent targets could be registered using the dGPS measurements.

To register the scans in Cyclone using the semi-permanent targets, the targets were identified and a point common to both scans obtained at t1 and t2 (e.g., on the nail head) was selected and tagged as a 'constraint.' Each constraint was given a unique name, which was shared in the t1 and t2 scans. A minimum of three constraints is required for registration. The mean RMS errors for scans registered in this way were 0.003 m for Inch and 0.012 m for Rossbehy. All registration errors at Inch were 0.01 m. The maximum error of registration was 0.045 m. The larger registration errors associated with the Rossbehy data may be due to small changes in the position of the posts in the field.

For the scans that did not share common registration markers from the semi-permanent targets, scans were registered using the dGPS coordinates of markers scanned in the field (either HDS targets or the semi-permanent markers). The dGPS coordinates of each of these markers were obtained on-site using a Trimble Pro-XH differential GPS. These data were downloaded and imported into Cyclone. The corresponding target locations were identified in the scanned data and then used to register the cloud to the imported dGPS data. As none of the semi-permanent markers were removed for the duration of the study at Inch, only scans at Rossbehy had to be registered in this way. The mean RMS error for scans registered using the dGPS was 0.109 m, considerably higher than for those registered using semi-permanent targets (0.012 m). This is likely to be due to the limited positional accuracy of the dGPS instrument used or the use of the Irish National Grid coordinate system rather than the GPS-optimised Irish Transverse Mercator. While the maximum registration error associated with this type of registration was 0.546 m, most registration errors (over 90%) were under 0.2 m.

VEGETATION FILTRATION

The dune surface at both field sites was obscured by dense marram (*Ammophila arenaria*) grass cover. Given the height of the vegetation (approx. 0.3–0.5 m) relative to the resolution of the scans (0.01 to 0.1 m), DEMs generated from raw point cloud data would not be representative of the true ground surface. As such, the vegetation had to be classified and removed from the scans.

Given the number of scans obtained during this research, an efficient automated approach to vegetation filtration was most desirable. Various approaches were considered and tested on subsets of data. These included:

(1) lowest points analysis—a technique whereby the lowest point within a grid cell of a specified size is considered to represent the ground;

(2) the use of reflected laser intensity distributions to differentiate between marram and the ground; and

(3) the use of the geometrical properties of points in the cloud to differentiate between the ground and vegetation.

The third option produced the best results. This was based on the work of Brodu and Lague (2012), who developed an algorithm called *CAractérisation de NUages de POints* (CANUPO) for classifying TLS point clouds in complex natural environments. The approach uses the geometrical properties of 3D scene elements across multiple scales to differentiate between them. It is based on the idea that at different scales, different elements within a 3D scene have different dimensionalities. Dimensionality is defined conceptually by Brodu and Lague (2012) as "how the cloud geometrically looks like at a given location and a given scale: whether it is more like a line (1D), a plane surface (2D), or whether points are distributed in the whole volume around the considered location (3D)." The quantitative measure of

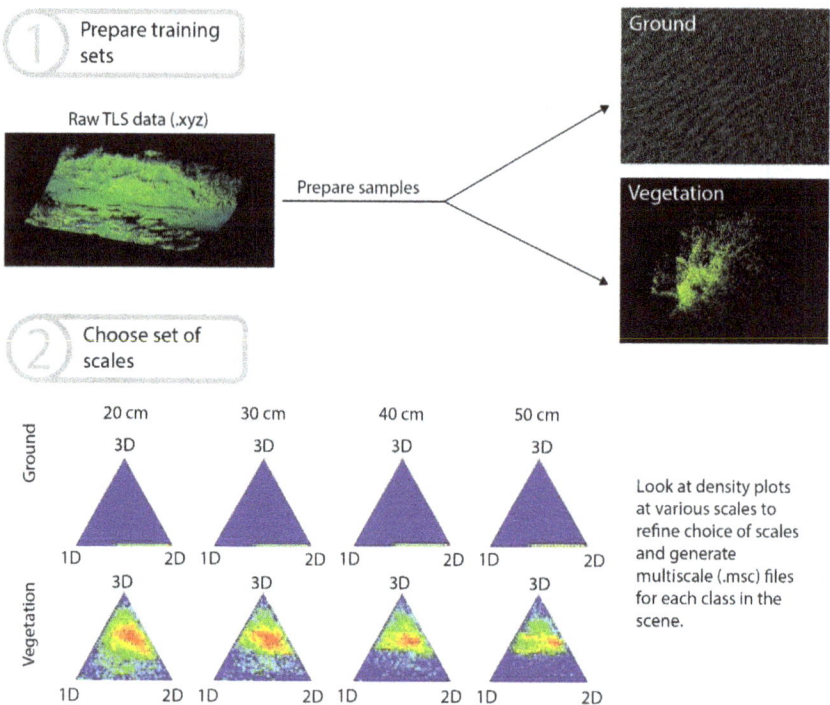

FIGURE 13.4 Workflow for vegetation classification with CANUPO. Classification is based on the 3D geometrical properties of the cloud at different scales. The first step is to prepare training sets—sample subsets of each feature within the cloud you wish to classify. Next, you must choose a set of scales over which the features have different geometrical properties. This is aided by visual inspection of density plots, which help to visualise the distribution of

(Continued)

dimensionality for each point is defined by the eigenvalues resulting from a principle component analysis (PCA) on the points. To give an example in the context of a vegetated dune environment, at very small scales, vegetation may appear more one or two dimensional (e.g., as stems and leaves), but at a larger scale, it will start to appear more three-dimensional (e.g., as a bush or tufts of grass). On the other hand, the ground surface may be more three-dimensional at a very small scale (e.g., ripples in the sand), but more two-dimensional at a larger scale (e.g., a beach). By exploiting these differences in dimensionality at different scales, it is possible to build unique signatures for identifying different categories of objects or elements within a scene.

Brodu and Lague (2012) developed this idea in the form of the CANUPO algorithm. CANUPO can be used to build site-specific classifiers, which can then be used to classify TLS point clouds. For this research, classifiers were built using the CANUPO plugin for CloudCompare, an open-source 3D point cloud and mesh processing software freely available from www.danielgm.net/cc/.

Figure 13.4 illustrates the workflow used in this research for classifier construction, which is described as follows. The first step was to prepare training sets, or 'examples', of each category—in this case, vegetation and ground. Training sets should be as representative as possible of each class and can include as many samples as necessary. Next, a relevant set of scale intervals must be specified for which

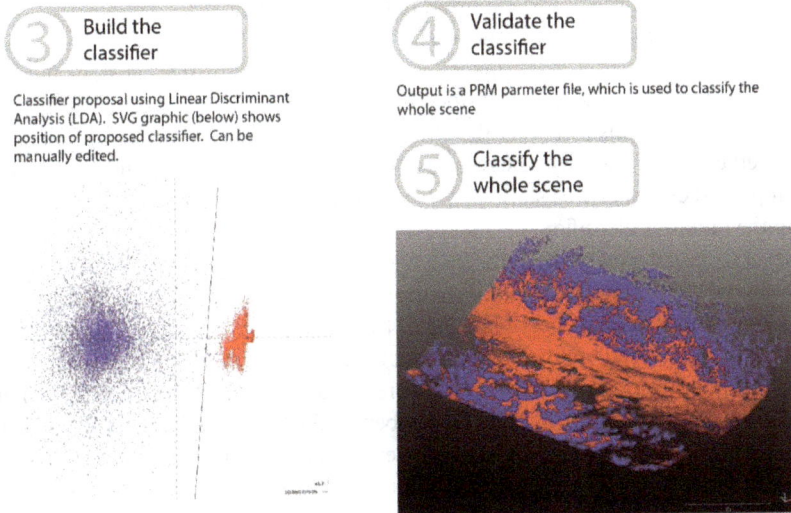

3 Build the classifier

Classifier proposal using Linear Discriminant Analysis (LDA). SVG graphic (below) shows position of proposed classifier. Can be manually edited.

4 Validate the classifier

Output is a PRM parmeter file, which is used to classify the whole scene

5 Classify the whole scene

FIGURE 13.4 (Continued)

dimensionality of the sample clouds. The more different the dimensionalities are at the different scales, the better for classification. Once the scales are selected, a classifier is proposed. The classifier should clearly separate the two sample data sets (based on the differences in their dimensionality at the different scales). Finally, on validation of the classifier, it can be applied to the whole scene. The output will be a cloud separated into the two classes, in this case, bare ground and vegetation. The bare ground cloud can then be exported for further analysis.

dimensionality between classes sufficiently differs. Local dimensionality is quantitatively defined in CANUPO using Principal Component Analysis (PCA), a statistical technique for finding patterns in data of high dimension. Initially, a 'best-guess' based on knowledge of the scene elements can be performed to aid in decision-making with regard to the identification of appropriate scale intervals. Further refinement can be achieved based upon visual analysis of density plots, triangular plots which aid in visualisation of the dimensionality, at each scale. Each corner of a density plot represents the tendency of the cloud to be 1D (lower left), 2D (lower right), or 3D (top). The plots are generated from the eigenvalue ratios calculated during PCA. Density plots for four scales are shown for the vegetation and ground classes of the Inch test data set in Figure 13.4 (under 'Step 2'). At all of these scales, the ground surface remains mostly two-dimensional, while the vegetation tends to become more three-dimensional as the scales increase. Based upon the dimensionality of the training sets at the specified scales, the algorithm then generates a probabilistic classifier by projecting the data in a plane of maximum separability between classes and then separating the classes in the plane. Figure 13.4 (under 'Step 3') shows an example of a proposed classifier. Classified points lie in the multiscale featurespace (red/blue) and the decision boundary (line separating the two) is generated using linear discriminant analysis (Koutroumbas and Theodoridis 2008). The classifier can also be generated using support vector machines, although both produce almost identical results. It is also possible to manually shift or tune the position of the decision boundary. Once the classifier is validated (Step 4) it can be applied to the entire scene (Step 5). The classes can be viewed, separated, and exported in CloudCompare.

The process of generating a successful classifier is often an iterative one. A good way to improve classifier performance is to identify false positives (e.g., ground classified as vegetation or vice-versa), include these as new training sets, and build the classifier again. In this research, the choice of best classifier was determined by testing whether or not the difference in residual error between the cloud filtered using the improved classifier was statistically significantly different from the cloud filtered using the previous classifier.

Several classifiers were built for Inch and Rossbehy. At Rossbehy, separate classifiers were built to deal with vegetation on the scarp (e.g., exposed roots) and on the beach (e.g., slump blocks and pioneer plants). For the scarp, scale intervals of 0.1 m from 0.7 m to 1.5 m were found to result in the best classifier performance. For the beach, scale intervals of 0.5 m from 0.5 m to 2.5 m were found to result in the best performance. For the lower resolution scans from December 2013 (10 cm) and January 2014 (15 cm), scale intervals of 1 m from 5 m to 20 m were found to result in best classifier performance for the beach. The classifier used on the scarps of the higher resolution scans performed well on these lower resolution scans so no new classifier was built for these. At Inch, scale intervals of 0.1 m from 0.1 m to 1 m resulted in the best classification.

Prior to applying these classifications to and filtering each data set, the raw scanned data had to be prepared. The first step was to remove erroneous data from the raw point clouds. Erroneous data can be generated from people or animals walking in front of the scanner, suspended sand or dust particles, or interference with direct sunlight. Erroneous data removal is fairly straightforward and can be done in either Cyclone or CloudCompare. Zones of poor or irregular resolution, usually at the far edges of the scans, and registration markers were also removed. Once the data were cleaned up, the classifiers were run on each data set, the vegetation was removed, and the bare-ground points were saved as ASCII text files for subsequent analysis.

GENERATION OF DEMS

Point clouds processed in CloudCompare were exported as ASCII text files and imported into ArcGIS v. 10.2 as xy data. The xy data were then exported to point shapefiles (along with corresponding z values) for further analysis. DEMs were generated from the point shapefiles using inverse distance weighting (IDW) with a variable search radius set to include 12 points (default). This method was chosen based on an evaluation of its performance against the empirical Bayesian kriging (EBK) and natural neighbour (NN) interpolations. Raster DEMs were exported at a resolution of 0.1 m for elevation and volumetric change analysis.

CHRONOTOPOGRAPHIC AND VOLUMETRIC CHANGE ANALYSIS OF TLS DATA

Chronotopographic (elevation change) and volumetric change analysis were also performed in ArcGIS. While a standard method of subtracting elevation (z) values between survey data obtained at t1 from survey data obtained at t2 could be applied to the Inch data, this was problematic at Rossbehy, where the scarp had shifted landward so much during the survey period that few scans actually overlapped in plan view. For example, during the short time between November 2012 and January 2013, the scarp had receded by more than 40 m. To address this issue, change analysis on the scarp at Rossbehy was performed in the horizontal, facing the scarp. The coordinate systems of the scarp clouds were translated in CloudCompare so that the y-axis, defined as perpendicular to the scarp, became the z-axis. This was undertaken prior to the generation of DEMs in ArcGIS. In some cases, parts of the beach at Rossbehy did overlap in plan form. In these cases, chronotopographic and volumetric change analysis was performed in the same way as on the Inch data.

The GIS workflow for generating elevation/distance change maps and calculating volumetric change between data from t1 to t2 is illustrated in the cartographic model shown in Figure 13.5 and outlined as follows:

1. Use minimum bounding geometry tool (set to convex hull) to create a perimeter around each raw TLS point shapefile. Edit as necessary to ensure that no gaps are present.
2. Use the intersect tool to create a new polygon, inside which DEMs from t1 and t2 will overlap.
3. Use the raster clip tool on exported raster DEMs with the polygon created in Step 2.
4. Use the raster calculator to subtract z values at t1 from z values at t2.
5. Format layout and export as cliff face/elevation change map.
6. Use the cut and fill tool to extract information about volumetric change (net gains and losses).

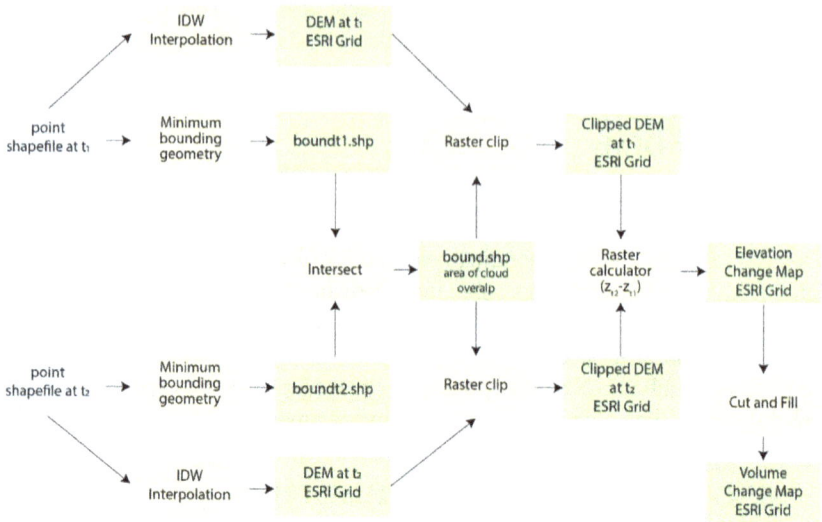

FIGURE 13.5 Cartographic model illustrating GIS workflow for generating elevation/dune scarp distance change maps and volume change maps.

Rates of volume change were calculated for each period using the formula of Young and Ashford (2006):

$$R_{vs} = V_s/(A \times T) \tag{13.1}$$

where:
R_{vs} = rate of volumetric change (m³ per m² per day)
V_s = volume change (m³)
A = areal extent of analysis (m²)
T = time between surveys (days)

The outputs of this analysis included elevation and scarp distance change maps; maps showing the location of net volumetric gains and losses; and information about mean elevation and scarp distance change, net volumetric gains and losses, and rates of volumetric change between surveys.

An assessment of error propagation due to vegetation filtering and DEM generation showed that the minimum levels of change detection ranged from ±0.05 m (±0.05 m³/m²) at Inch to ±0.44 m (±0.44 m³/m²) at Rossbehy.

FOREDUNE SCARP AND BEACH ELEVATION CHANGE

At Rossbehy, changes in the position of the foredune scarp and beach elevation were modelled separately due to the amount of dune recession (90+ m) observed over the study period. Changes to the position of the foredune scarp were modelled

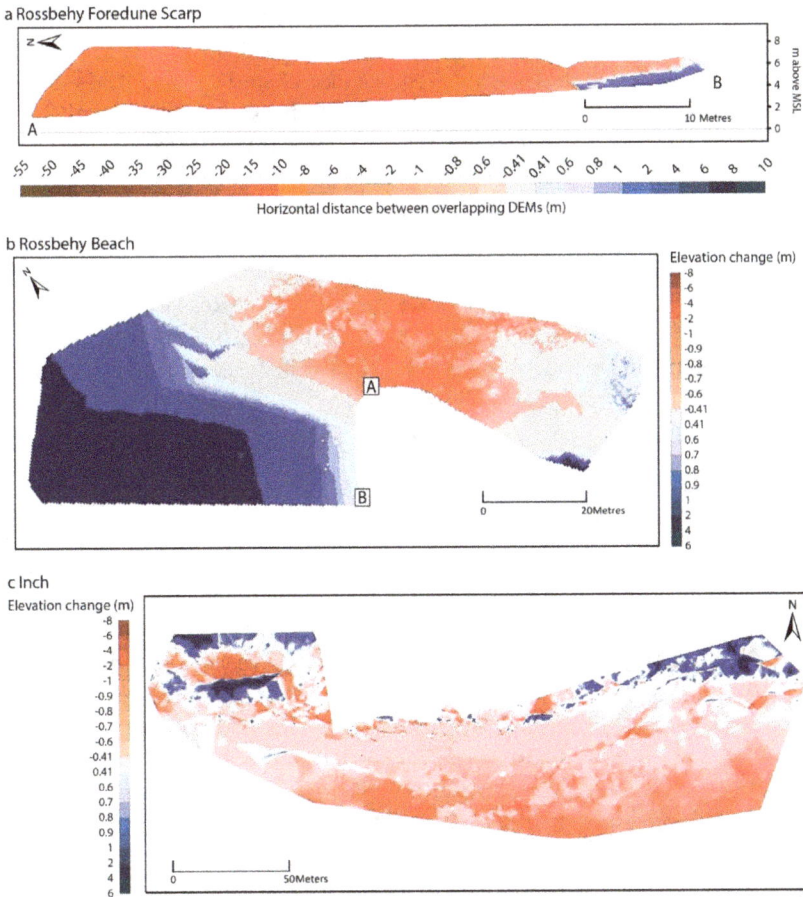

FIGURE 13.6 Selection of difference maps illustrating morphological change from summer to autumn 2012. (A) Rossbehy dune scarp distance change map; (B) Rossbehy beach elevation change map; (C) Inch beach/dune elevation change map. Due to the degree of shoreline recession at Rossbehy over the duration of the monitoring campaign (2012–2014), not all scans of the foredune overlapped in plan form. Scarp distance change, though, could be calculated by translating the cloud and calculating the distance of dune recession in the horizontal. In the example shown, part of the beach at Rossbehy overlapped in plan view, so beach elevation change could be mapped for this period. Locations "A" and "B" correspond to one another on the ground and illustrate the position of the foredune relative to the beach. By the end of the survey campaign in 2014, no part of the cloud overlapped in plan view. For each survey period, distance and elevation change was calculated using the raster calculator in ArcGIS, using DEMs at t_1 and t_2 as inputs.

perpendicular to the scarp in the horizontal direction (Δy), whereas changes in beach elevation were modelled in the vertical (Δz), as at Inch. As such, three separate maps were created for each period between surveys—one representing horizontal distance change in the position of the foredune scarp at Rossbehy; one representing beach elevation change at Rossbehy; and one representing beach and dune elevation change at Inch.

Figure 13.6 shows elevation and scarp distance change maps (DEMs of difference) for one of the survey periods for (a) Rossbehy scarp, (b) Rossbehy beach, and (c) Inch. Elevation change below the detectable limits (±0.41m for Rossbehy and ±0.05m for Inch) are shown in grey, and can be interpreted as relatively stable in terms of sediment movements. Clear pockets of erosion (red) and accretion (blue) are visible across the scarp face and beach. The total volumetric change was calculated for each period and used in later analyses to explore relationships between storm characteristics and volume changes. Net volume losses at Rossbehy were found to be significantly higher than at Inch for the study period, indicating an imbalance in sediment transfers to and from the sites. This imbalance is further explored in Kandrot (2016).

CONCLUSION

Terrestrial laser scanning provides a means to study the effects of storms and other geomorphologic drivers on beaches and coastal sand dunes. Issues around logistics in the field, scan registration, vegetation filtering, and DEM generation have been addressed. As technology rapidly advances, some of these issues are becoming less of a problem for practitioners. This work, however, reminds us of their relevance and provides practical guidance on how to effectively map the microscale geomorphology of coastal sedimentary systems and quantify morphological change. Such information is essential for the effective management of these extraordinary systems, which are increasingly coming under pressure due to the threats of sea-level rise and climate change. For more information about this work, see Kandrot (2016).

REFERENCES

Brock, John C, and Samuel J Purkis. 2009. "The Emerging Role of Lidar Remote Sensing in Coastal Research and Resource Management." *Journal of Coastal Research* 53: 1–5.
Brodu, Nicolas, and Dimitri Lague. 2012. "3D Terrestrial Lidar Data Classification of Complex Natural Scenes Using a Multi-Scale Dimensionality Criterion: Applications in Geomorphology." *ISPRS Journal of Photogrammetry and Remote Sensing* 68: 121–134.
Feagin, Rusty A, Amy M Williams, Sorin Popescu, Jared Stukey, and Robert A Washington-Allen. 2012. "The Use of Terrestrial Laser Scanning (TLS) in Dune Ecosystems: The Lessons Learned." *Journal of Coastal Research* 30 (1): 111–119.
Kandrot, Sarah. 2016. "The Monitoring and Modelling of the Impacts of Storms under Sea-Level Rise on a Breached Coastal Dune-Barrier System." University College Cork. https://cora.ucc.ie/handle/10468/3657.
Koutroumbas, Konstantinos, and Sergios Theodoridis. 2008. *Pattern Recognition*. 4th ed. Academic Press, Burlington, USA.
Leica Geosystems. 2006. *Leica ScanStation Product Specifications*. Heerbrugg, Switzerland.

Montreuil, Anne-Lise, Joanna Bullard, and Jim Chandler. 2013. "Detecting Seasonal Variations in Embryo Dune Morphology Using a Terrestrial Laser Scanner." *Journal of Coastal Research* 65 (sp2): 1313–1318.

Young, Adam P, and Scott A Ashford. 2006. "Application of Airborne LIDAR for Seacliff Volumetric Change and Beach-Sediment Budget Contributions." *Journal of Coastal Research* 22 (2): 307–318.

14 Creating a Virtual Reality Experience of Fingal's Cave, Isle of Staffa, Scotland

Victor Portela, Stuart Jeffrey, and Paul Chapman

INTRODUCTION: VR'S FALSE STARTS

Virtual Reality (VR) has had many false starts. In the early 1990s, films like *The Lawnmower Man* (IMDB 1992) inspired the press and the general public with a science-fiction VR storyline with 'cutting edge' computer graphics for the time. However, when the press descended on computer science departments around the world, they were disappointed with the bulky, low-resolution, high-latency head-mounted displays (HMDs) of the day that seemed to show no relationship to the exciting pre-rendered graphics from the film. Similarly, game companies that gambled on nascent VR products such as Nintendo's Virtual Boy (released in 1995) faced commercial failure.

GARTNER HYPE CYCLE

Looking back at the first wave of failed VR products, we are reminded of the work of the American researcher and futurist Roy Amara and his famous adage relating to forecasting the effects of technology:

> *We tend to overestimate the effect of a technology in the short run and underestimate the effect in the long run.*

(Ratcliffe 2016)

Amara's law is beautifully illustrated by the Gartner Hype Cycle (Figure 14.1). The respected consultancy firm Gartner provides this graphical representation annually to track the gradual adoption of a technology or product (Cearley and Burke 2018). Its Hype Cycle is divided into five phases:

The Technology Trigger. A new product/technology 'breaks through' via prototypes. There may be proof of concept demonstrations that can trigger significant media interest and publicity. At this stage, it is very rare for a usable product to exist.

DOI: 10.1201/9780429327575-14

The Peak of Inflated Expectations. The technology will be implemented by early adopters, and there will be a lot of publicity relating to its successful (and unsuccessful) implementation.

The Trough of Disillusionment. Interest dissolves as implementations of the technology fail to deliver. Investors continue to support the technology only if the problems can be addressed and the technology improved.

The Slope of Enlightenment. Second- and third-generation products emerge from companies and the technology starts to see more investment. More examples of how the technology can provide real returns on investment start to become understood.

The Plateau of Productivity. The technology is extensively implemented and its application is well-understood resulting in mainstream adoption. Standards start to arise for evaluating technology providers.

Figure 14.1 shows the Gartner Hype Cycle for AR and VR technologies for 2017 and 2018. Augmented Reality (AR) for 2017 (yellow circle) and 2018 (green circle) is positioned within the Trough of Disillusionment. In 2017, VR was positioned within the Slope of Enlightenment (yellow triangle). By 2019 both VR and AR were no longer visible on the curve. Why? Because Gartner no longer considers these technologies to be emerging. This is due to significant investment in these technologies and a growing demand and penetration into both industrial and home markets.

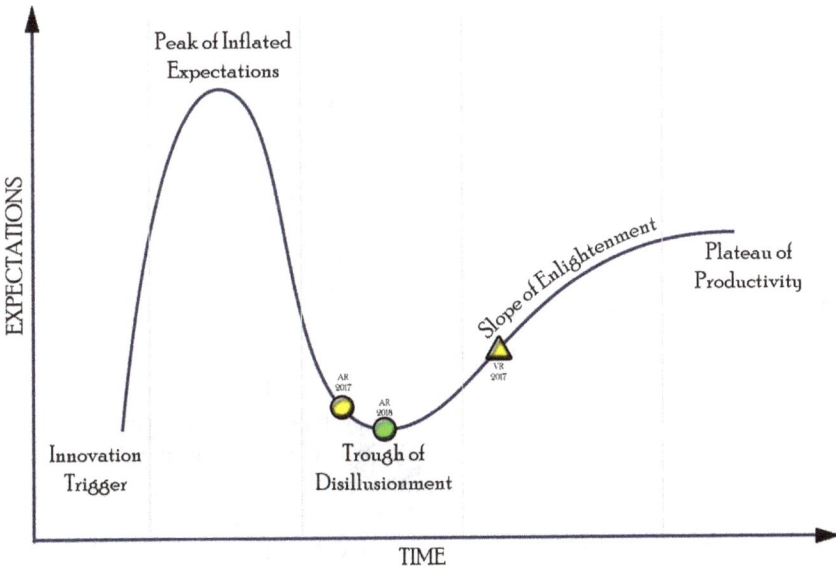

FIGURE 14.1 Gartner Hype Cycle 2017/2018—Augmented Reality (AR) and Virtual Reality (VR). In 2018, VR has matured sufficiently to move it off the emerging technology class of innovation profiles.

The rapid maturity of VR is further evidenced by the adoption of VR technology by local government in Scotland through the installation of headsets in primary and secondary schools, leading to a greatly enhanced student learning experience (BBC 2018). After VR's numerous false starts, it is satisfying to see the technology finally breaking through with a growing number of rapidly improving HMDs entering the market.

A RAPIDLY MATURING INDUSTRY

A major catalyst for VR escaping from the emerging technology curve was the pioneering work by Palmer Luckey, founder of Oculus, and the release of its first VR headset in 2013, The Oculus DK1. This device offered a much-improved resolution of 1280 x 800, a 90-degree field of view, and a low price of $300 USD, which was significantly cheaper than other HMDs of the day. Although it is now obsolete compared with the widely used Oculus Rift and HTC Vive, this headset was completely focused on developers, offering a development toolkit compatible with the widely used and well-established Unity3D and Unreal game engines as well as being compatible with all of the mainstream computer operating systems such as Windows, Mac, and Linux (Desai et al. 2014). The significant success amongst the developers and early adopters of the DK1 prompted new investment into Oculus (eventually purchased by Facebook for $2 billion USD (Dredge 2014) enabling the company to release an updated headset, the Oculus DK2, in mid-2014.

In summer 2014, another blue-chip company entered the VR scene but with a different approach. Google released the Cardboard (Cardboard Headset 2020), which, instead of focusing on developing an expensive headset, took advantage of the huge advances in mobile smartphones and used their displays for the VR graphics (Google Cardboard 2020). The case, which could be found online for $5 USD, made VR even more accessible for any kind of user or developer. As with the Oculus headsets, Google Cardboard released a development kit that was compatible with Unity3D, making the development of Google Cardboard apps accessible to all kinds of developers.

CONTEMPORARY VR HEADSETS

In 2016 the final versions of the Oculus Rift and HTC VIVE were released. Although by this time there were a number of HMDs on the market, these two devices have become the most widespread ones in use. They represented another step change over previously available devices. Both HMDs contained 2160 x 1200 resolution displays with a 100-degree field of view, with the Oculus Rift having a release price of $600 USD and $800 USD for the HTC VIVE (Greenwald 2016). These were not standalone headsets and needed to be connected to a computer providing all the computing power needed to drive the headsets. Thanks to the huge demands of the games industry, modern advances in computer graphics processor units resulted in a reduction in display latency, VR sickness, and cost that was the mainstay of early VR systems.

In 2020, both companies released the next generation of headsets offering improved features but maintaining affordability. The Oculus Rift S and Quest replaced the original Oculus Rift, and the Valve Index replaced the HTC VIVE, but even with the new wave of VR kits, the originals are still relevant due to backward compatibility and the maturity that the technology has reached over the past five to six years.

VR DEVELOPMENT TOOLS

The key feature that VR brings over other visualisation technologies is the immersion achieved by simulating the different senses of the human body. This allows the user to explore virtual worlds full of computer-generated interactive elements created by 3D modellers using tools such as ZBrush and 3DSMax. The modellers work closely with the programmers that load these assets into a videogame engine compatible with the chosen VR equipment. They then implement the features that will allow the end user to interact with the virtual environment (Figure 14.2).

There are several videogame engines compatible with VR technologies available today, but the two most popular in the VR development community are Unity3D and Unreal Engine. VR companies develop frameworks for their hardware that developers then import into their videogame engines greatly facilitating the development process.

A couple of examples of VR development frameworks are SteamVR, which is compatible with all of the Valve-related headsets (HTC Vive, Vive 2 or Valve Index), and Oculus SDK, compatible with all Oculus VR kits (Oculus Rift, Oculus Quest or Oculus Go). These are offered in software packages that are easily imported into any Unity project (Figure 14.3).

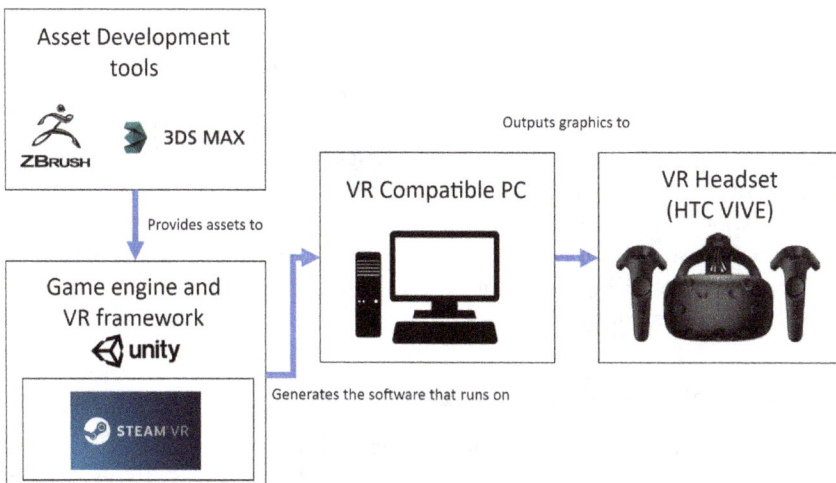

FIGURE 14.2 Typical hardware and software used in VR development (HTC VIVE example).

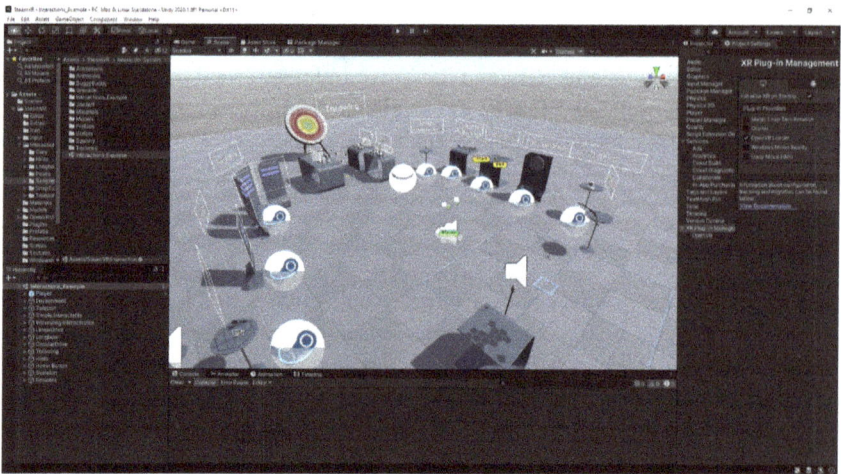

FIGURE 14.3   Unity3D loaded with the SteamVR framework.

INTEGRATING MULTIPLE INPUT DEVICES WITH VR APPLICATIONS USING UNITY

As mentioned earlier, VR relies on simulating different human senses in order to generate an immersive experience, but most VR kits are limited to simulating the sense of sight with the HMD and the sense of hearing with a pair of headphones.

In order to simulate other senses, such as the sense of touch with haptic feedback or to provide a different way to interact with the virtual elements other than the VR controller, the project must include additional hardware. Unity3D offers compatibility with many of these devices that have other development frameworks that can be integrated with any VR project. Some examples include the Phantom Omni (haptic interaction), the Leap Motion (hand tracking) and Microsoft Kinect (body tracking).

Another interesting feature that Unity3D offers to enhance VR projects is the addition of ambisonic sound. This technology allows the user to perceive sounds from any direction in 3D space even if they are using regular stereo headsets.

By using a combination of all the technologies mentioned here, VR projects can achieve a high degree of immersion, enabling more efficient ways for creating training, simulation, or teaching tools that can be used in many different fields outside the videogame industry. As an example, we consider the application of virtual reality to provide an immersive experience of Fingal's Cave, a National Nature Reserve and one of Scotland's major tourist draws.

FINGAL'S CAVE—A VR CASE STUDY

This section outlines a digital documentation and VR project case study derived from an ongoing multi-partner research project on the Isle of Staffa in Scotland, known as the Historical Archaeology Research Project on Staffa or HARPS. Beginning in 2014, HARPS is an interdisciplinary collaboration led by the Glasgow

Sschool of Art (School of Simulation and Visualisation) and the National Trust for Scotland, with partners from the University of Stirling, the University of Glasgow, and Spectrum Heritage. Off the west coast of Mull lies Staffa and its most well-known and striking feature, the large sea cave known as Fingal's Cave. The island itself is unoccupied and tiny, being only 1 km long and ½ km wide, but its unusual geology is quite apparent in the 40 m cliffs at its southern end. The geology of Staffa consists of a basement of volcanic tuff, overlain by a layer of Tertiary basaltic lava whose slow cooling resulted in strikingly regular basalt columns (Figure 14.4). Although unusual, this type of columnar basalt occurs in a number of places, including the 'Giant's Causeway' in Northern Ireland, but nowhere is it more spectacular than on Staffa. The regular geometric shapes are intriguing, and as late as the 19th century, scholarly discussions were still taking place as to whether the site was a human construction or entirely natural. Since first being brought to public attention by Joseph Banks in 1772 (Rauschenberg 1973), it rapidly became an essential destination for early tourists. Consequently, it has been the inspiration for works of music, art and literature by some of Europe's most important cultural figures, including Wordsworth, Turner, Verne and Hogg (for a fuller list see Eckstein1992). This intense interest in Staffa arose from 18th- and 19th-century romantic conceptions of the columnar basalt landscape and the rich folklore and oral traditions associated with the island (Michael 2007) and especially with the legend of Fhinn MacCool (and Macpherson's retelling of it (1772, Allen 1999)). Over time the island,

FIGURE 14.4   The entrance to Fingal's Cave from the sea, showing the striking columnar basalt column formations.

Source: Sian Jones

although natural, became a site 'burdened with culture' (Crane and Fletcher 2015). Despite its richly imagined past, peopled with heroes, sylphs, water nymphs and giants (McCulloch 1975), the reality was that Staffa remained a largely unknown quantity archaeologically until HARPS. Prior to 2014, the only archaeological work to have taken place on Staffa had been a walk-over survey for management purposes commissioned by the National Trust for Scotland in 1996 (Rees 1996), which noted numerous interesting, but undated, features.

HARPS has multiple research objectives: to uncover Staffa's material histories, the materiality of romantic travel and tourism, the remains of medieval and early modern settlement and farming and evidence of prehistoric activity. All of these avenues have borne fruit through archaeological excavation and analysis techniques. Evidence of a human presence on the island has been pushed back to the Neolithic, and a host of new material relating to early modern and tourist activity has been uncovered and recorded, including a large assemblage of historic graffiti (Alexander et al. 2018, 59, 2019, 45). However, in this case study, the focus is on the HARPS objectives of digitally documenting Staffa's iconic features and re-presenting the island to wider audiences by continuing the long-standing process of creative engagement with place through new media, specifically VR (Jeffrey 2015, 2018). With this objective in mind, a prototype VR environment was created that was ultimately used for multiple public engagement events, including during the 2019 Edinburgh Festival in partnership with the BBC (Jeffrey et al. 2019). The BBC also commissioned new music by the composer Aaron May (May 2020) in response to the VR, and this process is described in a BBC Radio 3 documentary and accompanying 360 Video (BBC 2019; Jeffrey et al. 2019).

The physical geometry of the VR environment was derived from both laser scans of Fingal's Cave (Figures 14.4 and 14.5) and the surrounding area and a much larger photogrammetric model of the coast of the whole island derived from many hundreds of high-resolution images taken from the sea (processed with Agisoft Metashape). These two data sets were combined to create an environment designed specifically to be experienced from sea level and to allow the user to travel right to the back of Fingal's Cave (which is normally inaccessible for visitors). It is important to note that although data sets exist that would allow a free exploration experience of the whole of the island (if using commercial LiDAR data for the non-coastal topography of the island's north end), it was the specific experience of approaching the cave from the sea that was considered most important for this model. Consequently, much data intrinsically valuable for archaeological and management purposes was not used in the VR model. Using the Unity games engine, described previously, dynamic wave animations were added to the model coastline as well as a flock of seabirds overhead, both of which were drawn from pre-existing commercial Unity asset libraries. The VR's skybox (weather simulation) and sun location were selected for their contribution to the atmosphere of the VR, and both are dark and somewhat foreboding. This reflects a key design decision underpinning the model's look and feel, which was to acknowledge that the physical experience of being in Fingal's Cave is not fully replicable in VR. It is very much an (extreme) multi-sensory experience, not simply visual, and the journey to the remote island is an intrinsic part of that experience. Therefore, rather than trying to create a potentially insipid or underwhelming version of visual reality, HARPS objective was

FIGURE 14.5   Laser scanning operations underway on west of Staffa in 2014.

Source: HARPS

to respond to the site by creating something entirely new in VR. Fundamentally they were not trying to recreate what it looks like to be in Fingal's Cave but evoke what it *feels* like to be there. This aligns with the long history of creative response around the cave where, over centuries, artists have sought ways to evoke the site without necessarily detailing its structure. Following from this decision, a further key aspect of the VR model was the acoustic experience; the strangeness of the sounds of the caves booming and crashing have long been noted and commented upon by Walter Scott, Jules Verne and many others, including Felix Mendelssohn, who famously took inspiration from the sea sweeping into the cave when writing the Fingal's Cave overture of his Hebridean Suite (Op.26, published 1832) (McCulloch 1975). In an effort to capture this property of the environment, HARPS conducted a sound sweep and directional microphone survey of the cave in 2014, producing digital convolution files that could be applied to the VR model for real-time spatialized auralisation (see Noble 2018 for a description of this process on Staffa). However, for the version of the VR to accompany the BBC documentary around the HARPS project, it was decided that an entirely new soundscape would be created, and the composer Aaron May was commissioned by the BBC to write an original piece of music. An interesting twist on the standard musical response to the landscape was that, unlike Mendelssohn, May never visited the island or the cave itself but responded entirely to being immersed in the VR experience that HARPS had created (Figure 14.6).

In keeping with the project's desire to reach as wide an audience as possible with Fingal's Cave VR, the final model was ported to multiple versions that could be delivered through various devices. The core version was designed for delivery via an

FIGURE 14.6 A screenshot of the VR, showing the exterior and the interior of the cave in VR created by the HARPS project for use with Aaron May's music.

Source: HARPS

HTC Vive. This version allowed free exploration and incorporated audio captured in the field as well as the real-time audio convolved from the acoustic model in 2014. It also allowed the user to speak, sing and shout into the cave space (note this required additional audio software at the time, in this instance Reaper). This version was also the one experienced by Aaron May. After he had composed his music, a single camera track was created that moved the user on a journey from the outside of the cave to the back of the cave and back again. Phases of the music are designed to complement the position of the user in the cave. A single camera track allowed the whole experience to be captured using spherical video, and this, in turn, allowed for the delivery of the experience via lightweight untethered devices such as the Oculus Go and even Google Cardboard.

The HARPS project digital documentation exercises have generated multiple data sets, with multiple uses, particularly regarding archaeological research and analysis. However, it was the project's desire to engage wider audiences, especially audiences who may not, for whatever reason, be able to visit the island in person that lead to our creation of the VR model. The driving force for the design of the model and the integration of music was a recognition that such uses of VR are not about replacing the experience of an in-person visit, but their ability to offer something different, but entirely complimentary to that visit.

CONCLUSION

Virtual Reality has finally come of age. We are finally able to move away from the gimmick installations and 'dragon demos' that have become ubiquitous at technology trade shows. VR's absence from Gartner's Hype Cycle since 2018 demonstrates a maturity of VR that is evidenced by industry's rapid and increasing adoption and application of the technology to real-world problems. VR has seen successful implementations in numerous disciplines including medical training, pharmaceutical manufacturing and the heritage sector. The next five years will see an explosion of lower cost and higher quality immersive helmet-mounted displays that will provide increases in resolution and field of view as well as improved ergonomics. After numerous false starts in VR, the race is finally on.

ACKNOWLEDGMENTS

The HARPS project would like to acknowledge the vital assistance of the National Trust for Scotland's London Members, the Society of Antiquaries of Scotland, Staffa Trips (of Iona) and Jack Kibble-White and Kate Bissell of the BBC.

REFERENCES

Alexander, D., Rhodes, D., Jeffrey, S., Jones, S., and Poller, T., 2018. *Staffa—Fingal's Cave Survey, Discovery and Excavation in Scotland New Series 18 2017.* Edinburgh: Archaeology Scotland.

Alexander, D., Rhodes, D., Jeffrey, S., Jones, S., and Poller, T., 2019. *Staffa—Excavation, Discovery and Excavation in Scotland New Series 19 2018.* Edinburgh: Archaeology Scotland.

Allen, P.M., 1999. *Fingal's Cave, the Poems of Ossian, and Celtic Christianity.* New York: SteinerBooks. ISBN: 0-826-1144-4

BBC, 2018. *East Renfrewshire Gives All Its Schools VR Headsets,* www.bbc.co.uk/news/uk-scotland-highlands-islands-43451583 (accessed January 29, 2021).

BBC, 2019. Virtually Melodic Cave. *Produced by Kate Bissel for BBC Radio Three.* BBC Sounds, www.bbc.co.uk/programmes/m00061lt (NB this link is permanent, but access to all BBC Sounds content is not available in all countries).

Cardboard Headset, 2020. https://arvr.google.com/intl/en_uk/cardboard/manufacturers/ (accessed May 29, 2020).

Cearley, D., and Burke, B., 2018. *Top 10 Strategic Technology Trends for 2018: A Gartner Trend Insight Report.* Stamford, CT: Gartner Publishing.

Crane, R., and Fletcher, L., 2015. Inspiration and Spectacle: The Case of Fingal's Cave in Nineteenth-Century Art and Literature. *Interdisciplinary Studies in Literature and Environment,* 22(4): 778–800.

Desai, P.R., Desai, P.N., Ajmera, K.D., and Mehta, K., 2014. A Review Paper on Oculus Rift—A Virtual Reality Headset. *arXiv preprint arXiv:1408.1173.*

Dredge, S., 2014. Facebook Closes Its $2bn Oculus Rift Acquisition. What Next? Online Article, *The Guardian.* www.theguardian.com/technology/2014/jul/22/facebook-oculus-rift-acquisition-virtual-reality (accessed May 29, 2020).

Eckstein, E., 1992. *Historic Visitors to Mull, Iona and Staffa.* London: Excalibur Press of London.

Google Cardboard Project, 2020. https://arvr.google.com/cardboard/ (accessed May 29, 2020).

Greenwald, W., 2016. *HTC Vive vs. Oculus Rift: VR Headset Head-to-Head*. Online Article, PC Mag, https://uk.pcmag.com/virtual-reality/77440/htc-vive-vs-oculus-rift-vr-headset-head-to-head (accessed May 29, 2020).

IMDB, 1992. *The Lawnmower Man—IMDb*, www.imdb.com/title/tt0104692/ (accessed January 29, 2021).

Jeffrey, S., 2015. Challenging Heritage Visualisation: Beauty, Aura and Democratisation. *Open Archaeology*, 1(1). http://doi.org/10.1515/opar-2015-0008

Jeffrey, S., 2018. Digital Heritage Objects: Authorship, Ownership and Engagement. In: *Authenticity and Cultural Heritage in the Age of 3D Digital Reproductions*, edited by P. Di Giuseppantonio Di Franco, F. Galeazzi, and V. Vassallo, pp. 49–56. McDonald Institute for Archaeological Research. Cambridge: University of Cambridge. https://doi.org/10.17863/CAM.27037

Jeffrey, S., Trench, J., Noble, S., Robertson, B., Breslin, R., May, A., Bissell, K., Kibble-White, J., Ramsay, S., Calderara, S., Simpson, J., Halkett, J., Rawlinson, A., Alexander, D., Jones, S., Rhodes, D., and Poller, T., 2019. *Edinburgh Festival BBC Virtual Reality Experience at Summerhall—Fingals's Cave 360 Video VR*. Edinburgh: Summerhall, 19–23 August 2019, http://radar.gsa.ac.uk/7005/ See also *Fingal's Cave and Aaron May*, www.youtube.com/watch?v=RHt6QIJI9cU&feature=youtu.be (YouTube, accessed June 2020).

Macpherson, J., 1772. *Fingal: An Ancient Epic Poem, in Six Books: Together with Several Other Poems, Composed by Ossian the Son of Fingal*. Translated from the Galic language, by James Macpherson. London.

May, A., 2020. www.aaronmay.co.uk/ (accessed June 2020).

McCulloch, D.B., 1975. *Staffa*, 4th ed., Newton Abbott: David and Charles (Holdings) Limited.

Michael, J.D., 2007. Ocean Meets Ossian: Staffa as a Romantic Symbol. *Romanticism*, 13(1): 1–14.

Noble, S.K., 2018. Fingal's Cave: The Integration of Real-Time Auralisation and 3D Models. *VIEW Journal of European Television History and Culture*, 7(14): 5–23. http://doi.org/10.18146/2213-0969.2018.jethc150

Ratcliffe, S., 2016. *Oxford Essential Quotations (4th ed.) Roy Amara 1925–2007*, American Futurologist. Oxford: Oxford University Press.

Rauschenberg, R.A., 1973. The Journals of Joseph Banks's Voyage up Great Britain's West Coast to Iceland and to the Orkney Isles July to October, 1772. *Proceedings of the American Philosophical Society*, 117(3): 186–226. www.jstor.org/stable/986542

Rees, T., 1996. *Survey of the Lands of the National Trust for Scotland on the Isle of Staffa*. AOC (Unpublished Archaeological Report).

Index

For Product Safety Concerns and Information please contact our EU
representative GPSR@taylorandfrancis.com
Taylor & Francis Verlag GmbH, Kaufingerstraße 24, 80331 München, Germany